£10-

Multiple Choice Questions in Physiology for the FRCS

Multiple Choice Questions in Physiology for the FRCS

James McDiarmid FRCS
Jason Bernard FRCS
Tamsin Greenwell
Alan Li FRCS
Nick Marshall FRCS
Chris Thurnell
Chris Stone FRCS

Edward Arnold
A member of the Hodder Headline Group
LONDON BOSTON SYDNEY AUCKLAND

First published in Great Britain 1995 by
Edward Arnold, a division of Hodder Headline PLC,
338 Euston Road, London NW1 3BH

Distributed in the Americas by Little, Brown and Company,
34 Beacon Street, Boston, MA 02108

British Library Cataloguing in Publication Data
A catalogue record for this book is available from the British Library

Library of Congress Cataloging-in-Publication Data
A catalog record for this book is available from the Library of Congress

ISBN 0-340-59434-9

1 2 3 4 5 95 96 97 98 99

Produced by Gray Publishing, Tunbridge Wells, Kent
Printed and bound in Great Britain by J. W. Arrowsmith Ltd, Bristol

Contents

PREFACE

The authors met whilst employed as anatomy demonstrators at the University of Manchester. After several barbecues and much revision they noticed the lack of a good MCQ revision book for the part A FRCS. Having passed the exam their attentions were turned to producing such a text which three years later has been published in the form of MCQs in anatomy, physiology and pathology for the FRCS.

The part FRCS has a reputation for being the toughest of all postgraduate examinations; whether this reputation is justified is debatable, however the scope of the applied basic sciences part of the examination is unarguably vast. With a recent change in the format of the exam even more emphasis has been placed on MCQs as a way of assessing candidates' knowledge. Both the style and the depth of question encountered in the primary have been recreated by seven authors who between them have passed the primary FRCS examinations of all three UK Colleges of Surgeons. It is also true that a large amount of the syllabuses of the FRCS and the MRCOG overlap and it is hoped that these books will also be found useful in this area. Detailed explanations to each answer are provided and it is hoped that readers will find these both unambiguous and educational.

The importance of focused revision cannot be overemphasized and these books are in no way supposed to take the place of the usual learning texts (listed below) and cadaveric dissection, indeed their roles lie in the final preparation for the exam once the bulk of reading has been done. We hope that these books will not only help in polishing exam technique but also clarify some of the important factual points that they initially found confusing.

Optimal MCQ examination technique varies considerably between candidates and the number of questions left unanswered in a negative marking MCQ test depends very much upon both one's core knowledge and the degree to which one is prepared to gamble if unsure of the answer. It is recommended that candidates experiment for themselves in order to elucidate what percentage of answered questions gives them the overall best mark; however the authors would tend to discourage excessive guessing.

The authors would like to extend their thanks to all those at the University of Manchester and elsewhere who aided their revision and helped to validate the content of these texts.

It only remains for the authors to wish you every success in your examination.

J. G. M. McDiarmid
C. A. Stone
December 1994

Recommended text

W. F. Ganong: *Review of Medical Physiology*, 16th Edition, Appleton and Lange, 1993.

Yes it's back, the book that you hoped as a medical student you'd never see again – although it's quite dry in parts you won't go wrong if you read it from cover to cover.

1 Sensory mechanisms and the central nervous system

1 With regard to the Pacinian corpuscle

A These are found only in the skin
B The afferent neurone's first node of Ranvier lies outside the corpuscle
C The corpuscle creates a receptor or generator potential in response to pressure
D The first node of Ranvier converts the graded response of the receptor into action potentials, the magnitude of which are proportional to that of the applied stimuli
E The generator potential is created by changes in nerve terminal permeability causing sodium ion influx

2 With regard to the sensory neurone

A Adaptation is said to occur when a maintained stimulus of constant strength causes action potential amplitude to decrease over a period of time
B Touch receptors show slow adaptation
C Receptors in the carotid sinus and muscle spindles adapt very slowly and incompletely
D Accommodation of the sensory neurone to the generator potential is partly responsible for adaptation
E The sensation evoked by stimulation of a receptor depends upon the specific part of the brain activated by the receptor's discrete sensory pathway

3 Regarding neural reflexes

A The knee jerk is initiated by stimulation of Golgi tendon organs in the quadriceps femoris tendon
B Muscle spindles comprise the intrafusal fibres of a muscle
C The ends of muscle spindles are non-contractile with all contraction occurring in the central portion
D Primary or annulospiral endings are the terminations of Ia rapidly conducting afferent nerve fibres and wrap around the central portion of the muscle spindle
E Gamma efferents provide the motor supply to muscle spindles

4 With regard to the reflexes

A A collateral from each Ia afferent synapses with an inhibitory interneurone that synapses in turn with a motor neurone supplying antagonistic muscles
B Stimulation of Golgi tendon organs causes muscle relaxation when their threshold is reached
C Clasp-knife rigidity in spastic hypertonic muscles is caused by an initial stretch reflex followed by an inverse stretch reflex
D Decreased gamma efferent discharge is found in clonus
E A crossed extensor response is found in the withdrawal reflex

5 With regard to sensory transmission

A Primary afferent neurones have their cell bodies in the substantia gelatinosa of the dorsal grey horn of the spinal cord
B Pain and temperature fibres ascend the spinal cord in the contralateral ventral spinothalamic tract
C The ventral and lateral spinothalamic tracts have a major input into the mesencephalic reticular formation
D Pain gating occurs by pre-synaptic inhibition of primary pain afferents in the substantia gelatinosa of the dorsal grey horns of the spinal cord
E Extraspinal tumours first cause loss of pain and temperature sensation in the lumbrosacral region

6 When the dorsal columns are destroyed

A The number of touch-sensitive areas on the skin is increased
B Vibration and proprioception are lost over the affected area
C The touch threshold is elevated
D Stimulus localization is impaired
E The effects are indistinguishable from those of destruction of the spinothalamic tracts

7 With regard to sensory transmission

A Diseases affecting the dorsal horns may lead to ataxia
B The skin has discrete warm- and cold-sensitive areas
C Thermoreceptors respond to the temperature gradient across the skin rather than the absolute temperature
D The synaptic neurotransmitter in pain fibres is substance P
E Pain causes an affective response via connections between the pain pathways and the thalamus

8 With regard to visceral pain

A Gastrointestinal pain from structures between the distal oesophagus and the sigmoid colon is sympathetically mediated
B Pain from the kidney, ureter and the superior part of the bladder is mediated via the parasympathetic nervous system
C Bronchial tree pain from the distal trachea to the diaphragm is conducted parasympathetically
D The cervix uteri and upper vagina have afferent pain fibres in the parasympathetic pelvic splanchnic nerves (S2–4)
E Pain transmitted from the testicle is sympathetic in nature and travels with the genitofemoral nerve

9 Regarding referral and inhibition of pain

A The radiation of anginal chest pain to the inner aspect of the left upper arm is an example of pain referral
B Embryonic migration of structures forms the basis for referral of pain
C The theory of convergence is based on subliminal fringe effects from neighbouring afferents lowering the threshold of spinothalamic neurones

D The theory of facilitation is based on the fact that visceral and somatic afferents synapse with the same spinothalamic neurone since there are far fewer spinothalamic neurones than there are afferent sensory nerve fibres
E Morphine acts to inhibit pain transmission both at the level of the afferent spinothalamic tract synapse in the substantia gelatinosa and at the level of the periaqueductal grey matter via the reticulospinal tract

10 Regarding vision
A Aqueous humour is produced by the ciliary body by a combination of diffusion and active transport
B Forward movement of the iris may cause open-angle glaucoma
C In the retina the rods and cones lie deep to the retinal blood vessels
D Müller (glial) cells bind together the neural elements of the retina
E The optic disc is the opthalmoscopic equivalent of the functional blind-spot

11 With regard to lesions of the visual pathways
A A lesion through the decussation of the optic chiasm leads to binasal heteronymous hemianopia
B A lesion of the left optic tract would cause a left nasal and right temporal homonymous hemianopia with macular sparing
C The outer segments of rods and cones are rich in mitochondria
D In retinitis pigmentosa there is a defect in the normal phagocytosis of rod outer segments in the retinal pigment epithelium
E Each foveal cell is connected individually to a single fibre in the optic nerve

12 Regarding vision
A Rods are specialized for night-time photopic vision
B Emmetropia is a term used to describe an abnormality of accommodation
C While gazing at a near object the ciliary muscle contracts
D In neurosyphilitic optic disease the pupillary light reflex is present but the accommodation reflex is absent
E Ultraviolet light of an appropriate wavelength causes rod and cone depolarization

13 With regard to hearing and equilibration
A The semicircular canals contain perilymph
B In the organ of Corti connections between the inner hair cells and the spiral ganglion are more frequent than those to outer hair cells
C Central auditory pathways pass from the spiral ganglion via the superior colliculus to the medial geniculate body and from there to the auditory cortex
D During rotational acceleration the cupulae of the semicircular canals move in a direction opposite to that of the acceleration
E The utricle and saccule contain endolymph

14 With regard to the inner ear

A Endolymph and cerebrospinal fluid have a very similar electrolyte composition
B Action potential formation in hair cells is due to potassium influx at their apices
C Normal conversation occurs at around the 20 dB range
D Maximum sensitivity of the human ear is in the 1000–4000 Hz range
E Tensor tympani and stapedius are both stimulated by loud sounds and have a reaction time of 40–120 ms

15 With regard to the inner ear

A High-pitched sounds resonate maximally near to the base of the cochlea
B When a constant speed of rotation is reached the cupula swings back to its upright position after a small delay
C Nystagmus, the fast phase of which is in a direction opposite to that of the rotation, is seen in the vestibulo-ocular reflex
D The saccule responds to horizontal acceleration
E Caloric stimulation of the ear can cause nystagmus, vertigo and nausea

16 Regarding olfaction

A In humans the olfactory mucous membrane has a surface area of 5 cm^2
B Each olfactory receptor cell synapses with an afferent neurone at its base
C Sustentacular cells in the olfactory mucous membrane secrete mucus in which substances must dissolve prior to their detection
D Pain fibres are found in the olfactory membrane
E Unilateral loss of smell in the absence of obvious nasal disease may indicate the presence of a meningioma

17 With regard to taste

A Taste buds have a diameter of 5–7 μm
B The epiglottis has taste buds in its mucosa
C Filiform papillae on the back of the tongue contain numerous taste buds
D The posterior third of the tongue has sensory taste fibres which travel with the seventh cranial nerve
E Captopril and penicillamine reduce taste sensation

18 Signs of cerebellar disease include

A Glabellar tap
B Dysmetria (past-pointing)
C Resting tremor
D Incoordination
E Dysdiadochokinesia

19 Transection of the spinal cord at C6 level will produce

A Immediate spastic paralysis of the trunk and extremities
B Diaphragmatic breathing only
C Urinary retention
D Ileus
E Hyporeflexic lower limbs

20 The following are essential in the diagnosis of brain death in a head-injured patient with a core temperature of 37°C in the absence of CNS depressant or muscle relaxing medications

A Absent gag and tracheal reflexes
B Lack of spontaneous respiration (with pCO_2 above 4.0 kPa)
C Lack of withdrawal reflex
D Absence of oculocaloric reflex
E Absent vERs

21 The following reduce the frequency of an adult alpha-type EEG

A Low blood glucose
B Increasing body temperature
C Reduced levels of adrenal glucocorticoids
D Increasing pCO_2
E Forced overbreathing

22 In diseases of the basal ganglia

A The basis of Wilson's disease is a low caeruloplasmin
B Chorea is associated with degeneration of the lenticular nucleus
C Hemibalismus is usually unilateral and due to contralateral subthalamic nuclear damage
D Clasp-knife rigidity is seen in Parkinson's disease
E Benzhexol acts as a dopamine agonist in anti-Parkinsonian therapy

23 Regarding the cerebellum

A A childhood medulloblastoma is characteristically midline
B Stimulation of the fastigial nucleus causes hypertension and increased ADH secretion
C Climbing fibres originate from the superior olivary nucleus
D Hypertonia is characteristic of cerebellar damage
E The basic cerebellar rhythm is 150–300 Hz

24 Regarding the autonomic nervous system

A At the junction between two sympathetic pre- and post-ganglionic neurones AChNIC (nicotinic acetylcholine) is the neurotransmitter
B Noradrenaline is the sympathetic post-synaptic neurotransmitter
C AChMUSC (muscarinic acetylcholine) is the post-synaptic parasympathetic neurotransmitter
D During fasting sympathetic nervous system discharges become more frequent
E Monoaminoxidase inhibitors increase circulating catecholamine levels

1 **A** = False **B** = False **C** = True **D** = False **E** = True

Pacinian corpuscles are found not only in the skin but also in the mesentery, muscles and joints. The first node of Ranvier lies within the corpuscle and converts the graded response of the receptor into action potentials, the frequency of which is proportional to the magnitude of the applied stimulus. The generator or receptor potential is created within the corpuscle via changes in sodium permeability at the myelinated nerve terminal permitting sodium influx in response to pressure. The generator potential is not propagated until it exceeds 10 mV.

2 **A** = False **B** = False **C** = True **D** = True **E** = True

Adaptation causes a reduction in action potential frequency rather than amplitude. Touch receptors adapt rapidly whereas receptors in the carotid sinus, muscle spindles and organs sensing cold, pain and lung inflation all adapt slowly and may adapt incompletely; these are known as tonic receptors. Accommodation of the sensory neurone to the generator potential and repositioning of the inner lamellae of the Pacinian corpuscle in response to a maintained stimulus are the causes of adaptation in this case. The 'doctrine of specific nerve energies' first enunciated by Müller states the contents of E above. For example if the afferent neurone of a Pacinian corpuscle in the hand is stimulated by pressure at the elbow or by irritation from a brachial plexus tumour the sensation evoked is one of touch; the corollary is true and is known as 'projection' – conscious sensations are referred to the location of the receptor – this is the basis of phantom limb pain in amputees.

3 **A** = False **B** = True **C** = False **D** = True **E** = True

The knee jerk is initiated by stimulation of muscle spindles in the quadriceps femoris muscle caused by passive stretching of the muscle. Muscle fibres are intrafusal and of either nuclear chain or nuclear bag variety, in both types all contraction occurs at the ends of the fibre. The statement in D is true; secondary or flowerspray endings are the termination of group II afferent neurones and attach to the ends of the intrafusal fibres. Gamma efferents are the motor supply to the muscle spindles and comprise 30% of the fibres in the ventral roots. Alpha–gamma linkage keeps the muscle spindles taut.

4 **A** = True **B** = True **C** = True **D** = False **E** = True

The statement in A describes reciprocal innervation; the inhibitory interneurone is known as a Golgi bottle neurone. When the stimulation threshold of Golgi tendon organs is reached muscle relaxation occurs; this is known as the inverse stretch reflex. Golgi tendon organs are a net-like group of knobbly nerve endings found among the fascicles of tendons. When they are adequately stimulated action potentials are transmitted via Ib afferents to create IPSPs (inhibitory post-synaptic potentials) on the motor neurones which supply the stretched muscle. Clasp-knife rigidity is indeed caused by an initial stretch reflex followed by an inverse stretch reflex as the stretching force is increased. Increased gamma efferent discharge is found in clonus, the muscle spindles are hyperactive making the muscle contract when briskly stimulated. This stops spindle discharge however, when the muscle relaxes again it is restimulated. In the withdrawal reflex a noxious stimulus causes flexion of the stimulated limb and extension of the contralateral limb – this is a polysynaptic crossed extensor response.

5 **A** = False **B** = False **C** = True **D** = True **E** = True

Primary afferent neurones have their cell bodies in the dorsal root ganglion. Pain and temperature fibres ascends the spinal cord in the contralateral lateral spinothalamic tract, gross touch and pressure ascend the cord in the contralateral ventral spinothalamic tract, the fine fibres ascending in the ipsilateral dorsal columns (fasciculus gracilis and cuneatus). The ventral and lateral spinothalamic tracts do input into the reticular formation thus maintaining the cortex in the alert state. The statement in D is true; there are various inputs the main ones being from the arcuate nucleus of the hypothalamus and contralaterals from the spinothalamic tracts via the periaqueductal grey to the nucleus retromedialis and from there to the substantia gelatinosa via the reticulospinal tracts. The statement in E is true; intraspinal tumours also first cause loss of pain and temperature sensation in higher segments due to the lamination of levels in the spinothalamic tracts and dorsal columns, respectively.

6 **A** = False **B** = True **C** = True **D** = True **E** = False

The number of touch-sensitive areas actually decreases due to loss of touch and pressure fibres. Vibration and proprioception are lost, the touch threshold is elevated and stimulus localization is impaired due to destruction of the fasciculi gracilis and cuneatus which carry touch and pressure fibres with temporal and spatial details. Destruction of spinothalamic tracts also causes an increase in the touch threshold and a decrease in the number of touch-sensitive areas on the skin, however the touch deficit is only slight and stimulus localization remains normal.

7 **A** = True **B** = True **C** = False **D** = True **E** = True

Diseases of the dorsal horns do lead to ataxia due to loss of proprioceptive fibres to the cerebellum. Mapping experiments have shown the presence of discrete warm- and cold-sensitive areas on the skin surface with 4–10 times as many cold spots as warm spots. Thermoreceptors actually respond to the absolute temperature and not the transcutaneous temperature gradient; these small A delta fibres have their nerve endings subcutaneously. Connections between the pain pathways and the thalamus are responsible for the affective component of pain – damage to the thalamus may lead to thalamic syndrome causing pain to be prolonged and severe.

8 **A** = True **B** = False **C** = False **D** = True **E** = False

Gastrointestinal pain between distal oesophagus and sigmoid colon is sympathetically transmitted via the greater, lesser and least splanchnic nerves; the vagus and glossopharyngeal nerves transmit parasympathetic pain proximal to the distal oesophagus and the parasympathetic pelvic splanchnic nerves (S2–4) mediate pain sensation distal to the sigmoid colon. The lesser splanchnic nerve and T11–L1 somatic nerves conduct afferent pain fibres from the kidney, the ureter and the superior part of the bladder. Pain from the distal trachea, parietal pleura and central diaphragm is conducted sympathetically via the brachial plexus, intercostal and phrenic nerves; bronchial tree pain originating proximal to the distal trachea is conducted parasympathetically in the superior laryngeal nerve and the upper thoracic vagal rami.

Pain transmitted from the testicles parasympathetically mediated with fibres travelling in the genitofemoral nerve (L1–2), the sacral nerves (S2–4) and the spermatic plexus (T10).

9 **A** = True **B** = True **C** = False **D** = False **E** = True

C describes the theory of facilitation and D the theory of convergence; afferents from visceral and somatic structures 'converge' on the same neurone thus confusing the cortex. The basis of referred pain is thought to be a combination of convergence and facilitation. Morphine does indeed act at the two levels described in E.

10 **A** = True **A** = False **C** = True **D** = True **E** = True

Forward movement of the iris to obstruct the anterior chamber angle causes angle-closure glaucoma. Open-angle glaucoma is caused by decreased permeability of the trabeculae at the canal of Schlemm. The rods and cones are the deepest layer of the retina. Müller (glial) cells bind together the neural elements of the retina and their processes form the internal and external limiting membranes. The optic disc is in fact the same as the blind-spot and represents the point at which the optic nerve enters the globe.

11 **A** = False **B** = False **C** = False **D** = True **E** = True

A lesion through the decussation of the optic chiasm would cause a bitemporal heteronymous hemianopia. A lesion of the left optic tract would cause a left nasal and right temporal homonymous hemianopia with no macular sparing – the pattern described in B would be seen in a lesion of the geniculocalcarine tract. Rod and cone outer segments contain pigment-filled saccules, the inner segments are rich in mitochondria and also contain the nucleus and the synaptic terminal. In retinitis pigmentosa debris accumulates between the receptors and the retinal pigment epithelium due to defective rod outer segment phagocytosis. The fovea contains only cones each one of which synapses with a single midget bipolar cell and thence to a ganglion cell which synapses directly with an optic nerve fibre permitting maximum resolution in this area.

12 **A** = False **B** = False **C** = True **D** = False **E** = False

Rods are specialized for scotopic vision in poor light, cones are for daytime photopic vision. Emmetropia occurs in optically normal eyes and means that parallel light rays are brought into sharp focus on the retina. Gazing at a near object requires ciliary muscle contraction to increase the convexity of the lens – the ciliary muscle has both radially and circumferentially orientated fibres. In the Argyll–Robertson pupil of neurosyphilis, tectal disease causes the pupillary light reflex to be absent but the accommodation reflex to persist. When stimulated by ultraviolet light rods and cones hyperpolarize, it is the amacrine cells that depolarize.

13 **A** = True **B** = True **C** = False **D** = True **E** = True

About 90–95% of neural connections in the organ of Corti are between the inner hair cells and the spiral ganglion. The central auditory pathway passes from the spiral ganglion via the inferior colliculus and thence to the medial geniculate body and the auditory cortex. During rotational acceleration both the perilymph and the cupulae of the semicircular canals move in the same direction as that of the applied acceleration.

14 **A** = False **B** = True **C** = False **D** = True **E** = True

Perilymph is very similar to CSF in respect of its electrolyte composition and endolymph is very similar to intracellular fluid. Movement of the hair cell stereocilia towards the kinocilium increases hair cell potassium permeability at their apices causing potassium influx which in turn activates calcium influx through voltage-gated calcium channels which in turn triggers neurotransmitter release. Whispering occurs at the 20 dB range, normal conversation is at around 60 dB. Reflex contraction of the stapedius and tensor tympani in response to a loud sound form the basis of the stapedial reflex which prevents hearing damage by overstimulation of the auditory receptors, however it is not rapid enough to protect against brief, intense stimulation due to explosions or gunshots.

15 **A** = True **B** = True **C** = False **D** = False **E** = True

Low-pitched sounds resonate maximally at the helicotrema the corollary of A above. In the vestibulo-ocular reflex the eyes move in a direction opposite to the direction of rotation – when the limit of movement is reached they quickly snap back to a new fixation point, hence the fast phase of the nystagmus is in the same direction as the rotation. The saccule responds to vertical acceleration and the utricle to horizontal acceleration. Caloric stimulation of the ear creates convection currents in the endolymph and hence nystagmus, vertigo and nausea.

16 **A** = True **B** = False **C** = True **D** = True **E** = True

The olfactory mucous membrane is located in the roof of the nasal cavity adjacent to the septum. There are no synapses at membrane level and the olfactory cells are themselves extensions of the first cranial nerve. Pain fibres are present in the olfactory membrane and are sensitive to irritants (e.g. chlorine and menthol), these fibres are conducted via the trigeminal nerve. A meningioma of the olfactory groove classically presents with unexplained distortion of smell. Loss of smell may also follow trauma, meningitis or occur idiopathically in the elderly.

17 **A** = False **B** = True **C** = False **D** = False **E** = True

Taste buds have a diameter of 50–70 μm. The mucosae of the epiglottis, pharynx, palate and tongue all contain taste buds. Filiform papillae do not contain taste buds, however fungiform and vallate papillae usually do. Taste fibres from the anterior two-thirds of the tongue travel with the chorda tympani branch of the facial nerve, while fibres from the posterior third travel with the glossopharyngeal nerve. Captopril and penicillamine produce hypogeusia due to sulphydryl groups in their structures.

18 **A** = False **B** = True **C** = False **D** = True **E** = True

Glabellar tap and resting tremor are signs of Parkinson's disease, intention tremor is seen in cerebellar disease. Dysdiadochokinesia is the inability to perform rapidly alternating opposite movements. In addition to B, D and E, other signs of cerebellar disease are an ataxic gait, scanning speech (slurred) and decomposition of movement.

19 **A** = False **B** = True **C** = True **D** = True **E** = False

Flaccid paralysis initially occurs, although some shoulder and elbow movement may be preserved. This flaccid paralysis may later proceed to spastic paralysis. Urinary retention remains unrestored and ileus occurs due to sympathectomy. Hyperreflexia in the lower limbs is initially seen.

20 **A** = True **B** = False **C** = True **D** = True **E** = False

Lack of spontaneous ventilation with a pCO_2 above 7.5 kPa is an essential criterion. Other signs like absent vERs, flat EEG, and carotid–jugular O_2 differential are helpful but not essential – the patient must be in the prestated conditions prior to testing for any of the above.

21 **A** = True **B** = False **C** = True **D** = True **E** = False

With regard to the EEG, infants have a 18–30 Hz beta rhythm, children normally can be observed to have 4–7 Hz theta rhythm and normal adults to have 8–12 Hz alpha rhythm. Low blood glucose, low body temperature, reduced levels of adrenal glucocorticoids and increasing pCO_2 all reduce alpha wave frequency. Forced overbreathing increases alpha wave frequency and is sometimes clinically used to unmask latent EEG abnormalities.

22 **A** = True **B** = False **C** = True **D** = False **E** = False

Caeruloplasmin is a copper-binding protein and its deficiency leads to copper intoxication and degeneration of the lenticular nucleus. Chorea is associated with degeneration of the caudate – lenticular lesions cause the slow writhing movements of athetosis. Hemibalismus is a striking sign involving flailing intense and violent involuntary movement of one side of the body, usually secondary to haemorrhage into the contralateral subthalamic nucleus. Lead-pipe and cogwheel rigidity are seen in Parkinson's disease. Benzhexol is an anticholinergic, bromocriptine is a dopamine agonist, both are used in anti-Parkinsonian therapy.

23 **A** = True **B** = True **C** = False **D** = False **E** = True

A childhood medulloblastoma would have all the characteristics of A above and usually presents with a broad-based staggering gait. Climbing fibres originate in the inferior olivary nucleus. Hypertonia is characteristic of cerebellar damage.

24 **A** = True **B** = True **C** = True **D** = False **E** = False

Fasting causes a reduction in the frequency of sympathetic discharges, post-prandially SNS discharge frequency is increased. MAOIs do not increase circulating catecholamine levels however they do increase catecholamine concentration in the brain.

2 CARDIOVASCULAR PHYSIOLOGY AND HOMEOSTASIS

1 Regarding the conduction system of the heart

A It is formed from modified nervous tissue
B The sinoatrial node is the natural cardiac pacemaker
C Vagal stimulation results in increased rate of sinoatrial node depolarization
D Atrioventricular nodal conduction is slowed by digoxin
E Conduction is the slowest through the atrioventricular node

2 In exercise

A The maximum heart rate seen in normal adults is approximately 230 beats per minute
B Increase in heart rate results in an equivalent reduction in duration of both systole and diastole
C Reduction in diastolic duration may result in cardiac ischaemia
D The radial artery pulse is felt 0.1 s after the peak of systolic ejection at rest
E A strong pulse indicates a high mean arterial pressure

3 Regarding cardiac auscultation

A The first heart sound is due to closure of the mitral and tricuspid valves at the start of atrial systole
B The presence of a third heart sound in young individuals warrants further investigation
C Left ventricular hypertrophy may be associated with a fourth heart sound
D A fourth heart sound may be due to delayed closure of the aortic valve
E The presence of a soft systolic murmur in children is a sign of significant underlying pathology

4 The following are required to measure cardiac output by the Fick principle

A Blood from the antecubital vein
B Blood from the pulmonary artery
C Blood from the radial artery
D Ventilation rate
E Indocyanine green

5 Oxygen consumption of the heart increases in

A Systemic hypertension
B Aortic stenosis
C Exercise
D Fear
E Angina pectoris

6 The following are commonly associated with oedema

A Kwashiorkor
B α-1-Antitrypsin deficiency
C Anaphylactic reaction
D Filiariasis
E Lymphoma

7 The following produce vasodilatation in most tissues

A Reduction in pO_2
B Reduction in $[H^+]$
C Reduction in $[K^+]$
D Increased production of kinins
E Increased [serotonin]

8 Following severe haemorrhage

A There is increased firing of atrial stretch receptors
B There is increased sympathetic discharge
C Increased aldosterone secretion precedes increased renin secretion
D There is increased baroreceptor discharge
E There is splanchnic venodilatation

9 Intracranial pressure

A Is increased by an infusion of crystalloid
B Is reduced by aircraft take-off
C Is reduced by downward head tilt at surgery
D Increases with increase in blood pressure
E Is raised by strenuous mental activity

10 During laparoscopic surgery

A The maximum insufflation pressure should be 40–50 mmHg
B 5–6 litre/min CO_2 should be insufflated
C Retroperitoneal haemorrhage can be controlled by immediately increasing insufflation pressure
D The patient should be hyperventilated
E Surgical emphysema is a recognized complication

11 With respect to an electrocardiogram (ECG)

A Depolarization moving towards an active electrode produces a negative deflection
B The T wave represents atrial repolarization
C A PR interval of 0.30 s is normal
D T wave inversion is a permanent sign of myocardial infarction
E At a $[K^+]$ of < 2.5 mmol/litre a peaked T wave is normally seen

12 Regarding the arterial pulse

A The radial pulse is the palpable forward motion of blood along the artery
B The dicrotic notch is palpable at the wrist
C The dicrotic notch occurs secondary to the snapping shut of the mitral valve
D A collapsing pulse is associated with mitral incompetence
E The velocity of the arterial pulse wave slows with advancing age

13 Cardiac output is depressed by

- A Inadequate post-operative analgaesia
- B Quinidine
- C Supraventricular tachycardia
- D Thyroxine.
- E Acute exposure to high altitude

14 The following are associated with an increase in central venous pressure in a normal individual

- A Intravenous infusion of 1 litre of crystalloid
- B Anaphylactic reaction
- C Constipation
- D Advanced dilative cardiomyopathy
- E Neoplastic obstruction of the superior vena cava

15 Regarding baroreceptors

- A They are high pressure circulation receptors only
- B They send afferent information to the brain via cranial nerves IX and XII
- C Impulse from these receptors result in a reflex reduction in heart rate
- D They do not discharge at blood pressure within the normal range
- E They reset in systemic hypertension so as to minimize rise in blood pressure

16 Regarding the cerebrospinal fluid (CSF)

- A The normal lumbar CSF pressure is less than 50 mmCSF
- B Reduces the net weight of the brain to less than 100 g
- C Removal via lumbar puncture may lead to pain which may be alleviated by intrathecal isotonic saline injection
- D Contains less protein than plasma
- E Contains 50% less glucose than plasma

17 During the cardiac cycle

- A The majority of ventricular filling is secondary to atrial contraction
- B The mitral and tricuspid valves close at the beginning of ventricular systole
- C The pulmonary valve opens when the right ventricular pressure exceeds that of the aorta
- D The end-diastolic ventricular volume is ~125 ml
- E Right atrial systole precedes left atrial contraction

18 Regarding the jugular venous pressure

- A The c wave is due to atrial contraction
- B The v wave is due to ventricular systole
- C The a wave is due to the bulging of the tricuspid valve into the atria during ventricular systole
- D Tricuspid valve insufficiency is associated with a giant c wave
- E Complete heart block is associated with a giant a wave

19 Cardiac output may be increased by

A Intravenous atropine
B Standing from the prone position
C Intravenous histamine
D Hypothyroidism
E Digoxin

20 Resistance to blood flow

A Increases with haemorrhage
B Is mainly within the venules
C Is increased by intravenous adrenaline
D Is reduced to < 10% of its normal resting value in a vessel by a tripling of vessel radius
E Is increased by polycythaemia

21 The Bezold–Jarisch (coronary) chemoreflex involves

A Apnoea following intravenous injection of serotonin
B Bradycardia
C Hypertension
D Tachypnoea
E Transient supraventricular tachycardia

22 During the cardiac cycle

A Atrial systole starts after the p wave of the electrocardiogram (ECG)
B Ventricular systole starts immediately after the s wave of the ECG
C Right ventricular systole precedes that of the left
D In expiration aortic valve closure precedes that of the pulmonary
E The ventricular ejection fraction is about 65% at rest

23 During exercise

A Heart rate and stroke volume increase proportionally
B Lack of parasympathetic innervation allows for increased cardiac output in the transplanted heart
C Blood flow is intermittent in skeletal muscle
D Oxygen consumption of skeletal muscle increases 1000-fold
E Cardiac output increases to >25 litre/min

24 Regarding the blood vessels

A The pressure in an ankle vein is ~90 mmHg whilst erect
B Greater than 50% of blood is stored in the systemic veins at rest
C Diastolic arterial pressure is measured at auscultation of the disappearance of Korotkov sounds
D Systolic blood pressure will increase in exercise to a greater degree than diastolic
E Systolic blood pressure increases to a greater degree than diastolic with age

25 Lymph

A Contains clotting factors
B Returns via the thoracic duct at a rate of 120 ml/h
C Removes over 30% of fluid filtered from arterioles
D Vessels contain valves
E Exists primarily for transport of fat from the gut

26 Lymphatic flow occurs secondary to

A Contraction of the walls of the large lymph ducts
B Contraction of the skeletal muscle in which the lymph vessels lie
C Contraction of the arteries accompanying the lymph vessels
D Contraction of the cardiac musculature
E Negative intrathoracic pressure during expiration

27 The following are true for skeletal muscle blood vessels

A Atropine produces vasodilatation
B Sympathectomy produces vasodilatation
C Noradrenaline infusion produces an increase in diameter
D Acetylcholine infusion produces vasoconstriction
E Histamine produces vasodilatation

1 **A** = False **B** = True **C** = False **D** = True **E** = True

The structures which compose the cardiac conduction system are the sinoatrial node, the atrioventricular node, the bundle of His and the Purkinje system. They are modified cardiac muscle. All component parts and the myocardium itself are capable of spontaneous discharge. The sinoatrial node has the most rapid rate of discharge and is therefore the natural cardiac pacemaker. The sinoatrial and the atrioventricular nodes receive both sympathetic and parasympathetic innervation. Vagal stimulation inhibits nodal depolarization – whilst sympathetic stimulation has the opposite effect. Other factors affecting the rate of discharge of nodal tissue are temperature and drugs. Digoxin exerts a depressant effect, especially upon the atrioventricular node. Conduction through this node is normally the slowest in the conducting system.

2 **A** = True **B** = False **C** = True **D** = True **E** = False

Theoretically the maximum rate of ventricular contraction is 400 per minute but in adults the atrioventricular node limits conduction so that the maximum rate seen is ~230 per minute. An increase in heart rate is accompanied by a reduction in duration of both systole and diastole but diastole is reduced to a much greater extent. Since diastole is when the majority of the myocardium obtains its blood supply from the coronary arteries (especially that of the left ventricle) reduction in duration may indeed result in cardiac ischaemia. The pulse (pressure wave) caused by forward movement of blood from the left ventricle into the aorta during left ventricular systole is felt at the radial artery at the wrist ~0.1 s after peak systolic ejection in the normal individual – the strength of the pulse being related to the pulse pressure and not the mean pressure.

3 **A** = False **B** = False **C** = True **D** = False **E** = False

The first heart sound is due to vibrations arising from the closure of the mitral and tricuspid valves at the start of ventricular systole. The second is caused by vibrations associated with aortic and pulmonary valve closure at the end of ventricular systole. The presence of a third heart sound is a normal finding in young adults whilst the finding of a fourth heart sound immediately before the first heart sound is rarely normal and may arise secondary to systemic hypertension, increased atrial pressure and ventricular hypertrophy. Many children have a soft systolic murmur and this does not indicate significant cardiac disease.

4 **A** = False **B** = True **C** = True **D** = False **E** = False

The Fick principle is that the amount of a substance taken up by an organ (or the whole body) per unit time is equivalent to the arterial concentration of the substance minus the venous concentration multiplied by the blood flow – and may only be applied to situations in which arterial blood is the only source of substance uptake. This principle is often used to measure cardiac output via the oxygen consumption of the body – requiring assays of oxygen consumption per unit time, arterial oxygen concentration (from any artery) and mixed venous oxygen concentration (from the pulmonary artery). Indocyanine green is used for cardiac output assay via the dye dilution technique.

5 **A** = True **B** = True **C** = True **D** = True **E** = False

Cardiac oxygen consumption is primarily determined by: (1) intramyocardial tension, (2) heart rate and (3) contractile state of myocardium. Increased oxygen consumption occurs with increased ventricular work. Ventricular work is the product of stroke volume and mean arterial pressure, therefore any factors which increase either will increase oxygen consumption. Systemic hypertension increases the after load and hence amount of work performed by the myocardium, as does aortic stenosis. Exercise increases oxygen consumption via an increase in heart rate and stroke volume, producing an increase in cardiac output. Fear produces an increase in sympathetic discharge which increases heart rate and hence oxygen consumption. Angina pectoris is a disease state arising from insufficiency of vascular (and hence oxygen supply) to the myocardium to meet normal demands.

6 **A** = True **B** = True **C** = True **D** = True **E** = False

Oedema is defined as the accumulation of abnormal amounts of fluid in interstitial tissue. Factors affecting interstitial fluid volume are

A Oncotic pressure
B Interstitial fluid pressure
C Capillary pressure
D Lymph flow
E Total extracellular fluid volume
F Capillary filtration coefficient

In Kwashiorkor an inadequate diet results in relative protein deficiency and hypoproteinaemia induced oedema, which is also the net result of hepatic cirrhosis of any cause including α-1-antitrypsin deficiency. Anaphylactic reaction results in the release of vasoactive substances v.v. histamine which increase capillary permeability and produce consequent oedema. Filiarial infestation obstructs lymphatic flow with resultant oedema.

7 **A** = True **B** = False **C** = False **D** = True **E** = False

Metabolic changes which produce vasodilatation in most tissues are

A Reduction in oxygen tension
B Reduction in pH
C Increase in temperature
D Increase in $[K^+]$
E Increase in [lactate]

Histamine and kinins produce vasodilatation in most injured tissues, whilst serotonin produces vasoconstriction.

8 **A** = False **B** = True **C** = False **D** = False **E** = False

After severe haemorrhage there is a reduction in extracellular fluid volume producing a reduction in discharge from all baroreceptors including atrial stretch receptor with a resultant increase in vasopressin secretion. Increased sympathetic discharge produces increased renin secretion which in turn results in increased aldosterone production. All the above aid in the production of generalized venoconstriction which diverts blood to the essential cardiac and cerebral circulation.

9 **A** = True **B** = True **C** = False **D** = True **E** = False

Intracranial pressure is proportional to venous pressure; an increase in one leads to a corresponding increase in the other. This provides a mechanism to compensate for arterial blood pressure changes at the level of the head. Factors increasing venous pressure v.v. a fluid load, raise intracranial pressure. Upward acceleration (as in aircraft take-off) results in a reduction in arterial blood pressure at head level, and hence a compensatory reduction in both venous pressure and intracranial pressure. The opposite is true for head tilt of the operating table. Physical signs of raised intracranial pressure are hypertension and bradycardia.

10 **A** = False **B** = True **C** = False **D** = True **E** = True

Maximum CO_2 insufflation pressure should be 10–15 mmHg. If retroperitoneal haemorrhage occurs then insufflation pressure should be immediately reduced, reducing caval compression and the patient should be immediately tilted head-down – it may be deemed necessary to proceed to laparotomy. The patient should be hyperventilated to 'blow off' absorbed CO_2.

11 **A** = False **B** = False **C** = False **D** = False **E** = False

An ECG represents the algebraic sum of action potentials of the myocardial fibres recorded extracellularly. Depolarizaation moving towards an active electrode produces a positive deflection whilst depolarization moving away from the active electrode causes a negative deflection. The ST segment and the T wave represent ventricular repolarization, atrial repolarization is not normally seen. The time range for a normal PR interval is 0.12–0.20. T wave inversion; a diagnostic ECG finding in myocardial infarction, completely resolves with time in many cases. Pathological Q waves are a more permanent ECG change. Hypokalaemia ($[K^+] < 2.5$ mmol/litre) is associated with a biphasic T wave and an ensuing positive U wave. A peaked T wave is associated with hyperkalaemia ($[K^+] > 6.6$ mmol/litre).

12 **A** = False **B** = False **C** = False **D** = False **E** = False

The pulse palpable at the radial artery at the wrist is due to the pressure wave set up by the forward motion of blood from the left ventricle into the aorta during systole – and not to the forward movement of the blood itself. The dicrotic notch is a small oscillation seen during the falling phase of the pulse wave, and is caused by the snapping shut of the aortic valve, and is impalpable at the wrist. With advancing age the arteries become increasingly rigid and this results in an increase in arterial pulse wave velocity. A collapsing or 'water hammer' pulse is a sign of aortic valve incompetence.

13 **A** = False **B** = True **C** = True **D** = False **E** = False

Cardiac output is the product of heart rate and stroke volume. Pain produces an increase in heart rate via an increase in sympathetic stimulation, and hence increases cardiac output. Quinidine is a type 1a antiarrhythmic drug which reduces heart rate and cardiac contractility (and hence stroke volume). Supraventricular tachycardia increases heart rate but also greatly reduces stroke volume. High altitude produces a physiological anaemia which is compensated for by an increase in stroke volume and hence cardiac output. Thyroxine is used for hormone replacement in individuals with hypothyroidism and will restore heart rate toward normal and therefore will increase cardiac output.

14 **A** = True **B** = False **C** = True **D** = True **E** = True

Central venous pressure is increased by:
(1) Increases in blood volume.
(2) Heart failure.
(3) Straining.
(4) Positive pressure ventilation.
(5) Superior vena cava obstruction.
Any factors resulting in the above will therefore produce an increase in central venous pressure.

15 **A** = False **B** = False **C** = True **D** = False **E** = False

Baroreceptors are stretch receptors situated in the walls of the right and left atria, the left ventricle and the pulmonary circulation. They send afferent impulses via cranial nerves IX and X (glossopharyngeal and vagus) – the reflex effects of which are (a) inhibition of vasoconstrictor tone and (b) excitation of cardiac vagal innervation. The net effects of (a) and (b) are vasodilatation, reduction in heart rate and hence a reduction in blood pressure. Baroreceptors are dynamic receptors, at normal blood pressure they discharge at a slow rate. This rate of discharge increases with increase in blood pressure and falls with reduction in blood pressure. Chronically elevated blood pressure results in a relatively rapid resetting of these receptors in order to maintain this elevated blood pressure as the new 'normal'.

16 **A** = False **B** = True **C** = True **D** = True **E** = False

Lumbar CSF pressure is normally 70–180 mmCSF. CSF forms a protective water bed for the brain. The effective weight of the average adult brain is 1500 g in air and 50 g in CSF. Removal of CSF at lumbar puncture reduces this 'water bath' effect leading to headache which may indeed be alleviated by intrathecal injection of isotonic sterile saline. CSF differs from plasma in several notable respects (ratio cf. plasma):
(1) 2/3 [K^+].
(2) 1/2 [Ca^{2+}].
(3) ↓↓↓ greatly reduced [protein].
(4) 2/3 [glucose].
(5) ↑ pCO_2.
(6) ↓ pH.

17 **A** = False **B** = True **C** = False **D** = True **E** = True

Greater than 70% of ventricular filling occurs passively during diastole with atrial systole only contributing a small percentage. The mitral and tricuspid valves close at the start of systole. The pulmonary valve opens when the right ventricular pressure exceeds that of the pulmonary artery (~10 mmHg) and the aortic valve opens when the pressure in the left ventricle exceeds that of the aorta (~80 mmHg). The end-diastolic ventricular volume is around 125 ml. There is asynchronicity of events in the cardiac cycle – and right atrial systole precedes left atrial systole.

18 **A** = False **B** = False **C** = False **D** = True **E** = True

There are three waves of note seen on inspection of the jugular venous pulse. The a wave is due to atrial systole. The c wave occurs as a result of the bulging of the tricuspid valve into the right atrium during the isovolumetric phase of ventricular systole. The v wave occurs secondary to passive venous filling of the atria during diastole prior to opening of the tricuspid valve. Tricuspid valve insufficiency is associated with giant c waves. Complete heart block is associated with giant a waves (cannon waves).

19 **A** = True **B** = False **C** = True **D** = False **E** = True

Cardiac output is the product of heart rate and stroke volume. Any factors increasing either will therefore increase cardiac output. Under resting conditions heart rate is under parasympathetic inhibition. Atropine blocks parasympathetic discharge and will therefore increase heart rate and cardiac output. Histamine causes vasodilatation and will produce an increase in cardiac output via an increase in stroke volume. Hypothyroidism reduces resting heart rate and hence cardiac output. Standing from the prone position produces a fall in venous return and hence stroke volume. Digoxin has a positive inotropic effect, increasing myocardial contractility and hence stroke volume.

20 **A** = True **B** = False **C** = True **D** = True **E** = True

According to the Poiseuille–Hagen formula:

$$R = \eta L/r^4,$$

where R is the resistance, η is the viscosity, L is the length of the tube and r is the radius of tube. Therefore $R \propto 1/r^4$ and hence an increase in $r \times 3$ results in a reduction in R by a factor of 81 (e.g. much less than 10%). Following haemorrhage there is an increase in resistance to blood flow secondary to reflex constriction of blood vessels. Vascular resistance is mainly sited within the arterioles. Intravenous adrenaline provokes sympathomimetic arteriolar constriction and increased vascular resistance. Resistance to blood flow depends mainly upon the diameter of the blood vessel but also to a lesser extent upon the viscosity of the blood. Polycythaemia produces an increase in viscosity and hence an increase in resistance to flow.

21 **A** = False **B** = True **C** = False **D** = True **E** = False

The Bezold–Jarisch reflex is secondary to excitement of left ventricular receptors by injection of a variety of substances including serotonin into the arterial supply of the left ventricle, and is a response of apnoea followed by tachypnoea, hypotension and bradycardia.

22 **A** = True **B** = False **C** = False **D** = False **E** = True

Atrial systole follows the p wave of the ECG and ventricular systole commences near the end of the r wave and continues until after the t wave. Whilst right atrial systole precedes left atrial systole, right ventricular systole lags that of the left ventricle. In expiration the aortic and pulmonary valves close simultaneously and it is in inspiration that the aortic valve closes before the pulmonary valve. The stroke volume at rest is 70–80 ml and the resting end-diastolic ventricular volume is 125 ml. Therefore the ejection fraction is approximately 65%.

23 **A** = False **B** = False **C** = True **D** = False **E** = True

In the normal individual exercise increases sympathetic discharge which produces a larger increase in heart rate then stroke volume. Patients with transplanted and hence denervated hearts are able to achieve an increase in cardiac output during exercise by increasing cardiac return (the Frank–Starling mechanism). During exercise skeletal muscle contraction results in the cessation of blood flow through the muscle when >70% of maximum muscle tension has developed. Oxygen consumption increases 100-fold and blood flow 30-fold during exercise producing a maximal cardiac output of ~35 litres/min.

24 **A** = True **B** = True **C** = False **D** = True **E** = True

Peripheral venous pressure is affected by gravity and increases by ~0.77 mmHg/cm below the right atrium and is ~90 mmHg in an ankle vein whilst erect. The veins are the capacitance vessels and store >50% of the circulating blood volume at rest. In the U.K., diastolic blood pressure is measured as the pressure at which Korotkov sounds become muffled. Both systolic and diastolic blood pressures increase with age – but systolic increases to a greater degree. Exercise increases heart rate and stroke volume however it also reduces total peripheral resistance via vasodilatation and therefore systolic blood pressure increases moderately whilst diastolic remains constant or even falls slightly. With increasing age systolic blood pressure increases more than diastolic.

25 **A** = True **B** = True **C** = False **D** = True **E** = False

The lymphatic system is primarily concerned with the removal of protein and fluid from the interstitial space. Ninety percent of the fluid filtered at the arteriolar end of the capillary network later returns into venules and the remainder is removed by lymph.

26 **A** = True **B** = True **C** = False **D** = False **E** = False

Lymphatic flow occurs secondary to:

(1) Contraction of surrounding skeletal muscle.
(2) Negative intrathoracic pressure during inspiration.
(3) Suction effect of blood flow through the large veins into which the lymph drains.
(4) Rhythmic contraction of the wall of the large lymphatic ducts.
(5) Pulsations of adjacent arteries.

27 **A** = False **B** = True **C** = False **D** = False **E** = True

Vasodilatation in skeletal muscle vessels occurs secondary to:

(1) $\downarrow$ pO_2.
(2) $\downarrow$ pH.
(3) $\uparrow$ pCO_2.
(4) $\uparrow$ Osmolality.
(5) $\uparrow$ Temperature.
(6) $\uparrow$ $[K^+]$.
(7) $\uparrow$ [lactate].
(8) Histamine.

Skeletal vasculature is under resting sympathetic vasoconstrictor tone and sympathectomy produces vasodilatation. Skeletal muscle vasculature also receives innervation from the cholinergic sympathetic vasodilator system. Acetylcholine infusion will therefore produce vasodilatation whilst atropine will not.

3 PULMONARY FUNCTION AND RESPIRATORY HOMEOSTASIS

1 Regarding spirometry and respiratory volumes

A FEV_1 is greater than 3.5 litres in the adult
B Residual volume of 1 litre may be determined by spirometry
C Expiratory reserve volume is measured by active not passive effort
D Maximal respiratory minute volume is function of vital capacity only
E Alveolar exchange is 150 ml per breath

2 Intrapleural pressures are

A Never positive at tidal volumes in the normal adult
B Lower at the apices than the bases
C Greater with an open than a tension pneumothorax
D More negative in inspiration with greater airway resistance
E When rising cause a positive intra-alveolar pressure

3 Physiological dead space

A Is increased dead space
B Is greater than anatomical dead space in health and disease
C Can be calculated from tidal volume, inspired and expired pCO_2
D Is unchanged with rapid respiratory rate if the minute volume remains at 6 litres/min
E Alone accounts for differences in alveolar and expired air composition

4 Alveolar surfactant

A Concentration is dependent on lung volume
B Depletion will decrease compliance
C Aids ventilation and diffusion
D Can cause atelectasis
E Depletion results in pulmonary oedema in hyaline membane disease

5 Pulmonary compliance is

A Maximal during tidal respiration
B Greater at the apices
C Dependent solely on thoracic cage and lung elasticity
D Increased in congestive cardiac failure
E Altered with altered airway conductance

6 Physiological right to left shunt

A Causes a decrease in arterial pO_2 of 2 mmHg
B Occurs in the heart
C Is greater in the newborn than the foetus
D Occurs between pulmonary arteries and bronchial veins
E Lowers the ventilation/perfusion ratio

7 Pulmonary vascular resistance

A Decreases with hypercapnia
B Increases with acidaemia
C Increases with both circulating adrenaline and increased sympathetic discharge
D Is maximal in the pulmonary veins
E Increases with histamine release by causing pulmonary venule constriction

8 The ventilation/perfusion ratio

A Increases from apex to base in the erect position
B Decreases with increased physiological dead space
C Is decreased in exercise
D Pulmonary embolus causes a reduction in both ventilation and perfusion
E Is increased by haemorrhage

9 Oxygen dissociation curve is shifted to the right by

A Decrease 2,3-dpg
B Altitude
C Rise in pCO_2
D Increase in pH
E B-thalassaemia

10 In passing through the peripheral tissues, haemoglobin

A Changes molecular configuration
B Iron is reduced from ferrous to ferric form
C More readily forms carbaminohaemoglobin
D Buffers less
E Contains more carbonic anhydrase

11 Carbon monoxide

A Poisoning is diagnosed by arterial blood gas sampling
B Has greater affinity than carbon dioxide for haemoglobin
C Causes anaemic hypoxia
D Shifts the O_2 dissociation curve to the left
E Dissociation from haemoglobin is unaffected by 100% oxygen at 1 atmosphere pressure

12 Carbon dioxide is

A Carried mainly in plasma
B Carried mainly as dissolved CO_2
C Hydrated before crossing red cell membrane
D Converted into bicarbonate in plasma
E Combined with haemoglobin

13 Carbon dioxide

A Penetrates the blood–brain barrier slowly
B Is irreversibly hydrated by carbonic anhydrase

C Is more soluble than oxygen
D Increases proportionately with decreased ventilation
E Is more diffusible than oxygen

14 Cyanosis is often present with

A Polycythaemia
B Congestive cardiac failure
C Tension pneumothorax
D Chronic obstructive airways disease without dyspnoea
E Ventricular septal defect

15 Hypercarbia occurs in the following

A Head injury with Glasgow coma score of less than 8
B Prolonged vomiting
C Compensated and uncompensated respiratory acidosis
D Salicylate toxicity
E Pulmonary fibrosis

16 Arterial pCO_2

A Rise causes increased bicarbonate excretion
B Normal value is 45–56 mmHg
C Is invariably elevated in chronic bronchitis
D Stimulates both peripheral and central chemoreceptors
E Decreases at altitude

17 Hyperventilation causes the following

A Reduced total blood Ca^{2+} to cause tetany
B Low plasma HCO_3^-
C Alkalosis to pH 7.5–7.6
D Shift in O_2 dissociation curve to left
E Elevated K^+

18 Breathing 5% CO_2 causes

A Hyperventilation
B Headache
C Sleepiness
D Bradycardia
E Pulmonary vasoconstriction

19 The following are correct

A Impaired alveolo-capillary diffusion may cause hypoxaemia with hypocarbia
B Arterio-venous shunting improves markedly with inspired 100% O_2
C Ventilation/perfusion inequality causes hypoxaemia, hypercarbia and acidosis
D Acidosis and hyperventilation result from uraemia
E Hypoventilatory hypercapnia is proportionately corrected by increased ventilation

20 Regarding pneumothoraces the following are correct

- A Air moves preferentially through a chest wound 2/3 tracheal diameter
- B Increases vagal afferent stimulation
- C Tension pneumothorax is diagnosed radiologically
- D Closed pneumothorax increases pulmonary vascular resistance
- E May cause cardiovascular collapse

21 Chronic bronchitis and emphysema are associated with

- A Reduced total lung capacity and functional residual capacity
- B Reduced expiratory time
- C Maximal airway resistance in expiration
- D Destruction of pulmonary capillaries with cor pulmonale
- E Increased diffusible lung surface area

22 The following are true

- A Pulmonary embolism increases dead space
- B Pulmonary embolism decreases shunting
- C Pulmonary oedema reduces ventilation/perfusion ratio
- D Pulmonary oedema increases airway resistance
- E FEV_1:FVC ratio may increase in pulmonary fibrosis

23 Major abdominal surgery is post-operatively associated with

- A Increased tidal volume
- B Impaired cough reflex
- C Reduced arterial pO_2
- D Decreased respiratory minute volume with opiate analgesia
- E Decreased likelihood of atelectasis

24 Respiratory centre is stimulated by

- A Haemorrhage with reduced arterial pulse pressure
- B Increased carotid sinus discharge
- C Lung inflation receptor discharge
- D Carbon monoxide poisoning
- E Cerebral injury

25 Carotid body chemoreceptors are stimulated by

- A Haemoglobin of less than 10 g/dl
- B Elevated CO_2 more than reduced O_2
- C Partial pressure and total content of arterial O_2
- D Increased blood flow
- E Decreased H^+ ion concentration

1 **A** = True **B** = False **C** = True **D** = False **E** = False

Vital capacity for a 70 kg male is 4.5 litres and forced expiratory volume in 1 s is greater than 80% of this. Calculation of residual volume requires the use of helium in a closed circuit spirometer and determination of its dilution as it is not appreciably taken up. Expiratory reserve volume remains at end tidal expiration, a largely passive phenomenon. Hence active effort is required to expel the remaining capacity. Maximal breathing capacity is also dependent on respiratory rate. Anatomical dead space of 150 ml, i.e. gas in the conducting airways not equilibrating with pulmonary blood, is a proportion of each tidal breath of 500 ml.

2 **A** = True **B** = False **C** = False **D** = True **E** = True

At tidal volumes, negative pressures maintain open airways even at end expiration where recoil pressures between lung and chest wall balance with a pressure of –2.5 mmHg. This becomes positive with forced expiration to effect airway closure in the bases first as the weight of the lung results in less negative, i.e. lower intrapleural pressures here. Open pneumothorax results in equilibration with atmospheric pressure whereas a tension pneumothorax raises the pressure beyond this. Increased airway resistance requires greater inspiratory effort to generate greater intrapleural pressures to move the same amount of air. Any rise in pressure results in a positive intrapulmonary pressure and movement of air outwards.

3 **A** = False **B** = False **C** = False **D** = False **E** = True

Physiological and anatomical dead space are identical in healthy individuals. It may be calculated from the Bohr equation requiring expiratory and alveolar pCO_2 and tidal volume. Inspired pCO_2 is negligible. If respiratory minute volume is unchanged with greater respiratory rate, the tidal volume must be less and the dead space a proportionately greater amount of this.

4 **A** = True **B** = True **C** = False **D** = True **E** =

Greater concentration of surfactant exists at smaller volumes, thus reducing surface tension with decreasing alveolar radius. This minimizes the lowering of distending pressure, in turn required to keep alveoli open (Laplace's law $P = 2T/r$). Deficiency causes alveolar collapse, atelectasis and unopposed surface tension increase, encouraging fluid transudation from blood.

5 **A** = True **B** = False **C** = False **D** = False **E** = True

Compliance, the change in lung volume per unit change in airway pressure, is a function of lung and thoracic cage elasticity, surface tension and stretch against inelastic components. It is not purely linear in function and degree of compliance varies with varying lung volumes with the steepest part of the relaxation pressure curve occurring during tidal respiration. Apical alveoli exist at greater starting volumes where proportionately less expansion occurs compared with basal alveoli. Resistance, the inverse of conductance, also varies during the respiratory cycle to produce a hysteresis loop in the pressure volume curve. Pulmonary oedema associated with cardiac failure increases the stiffness of the lung.

6 **A** = True **B** = True **C** = False **D** = False **E** = True

Alveolar pO_2 is 100 mmHg, that in pulmonary capillaries is approximately 2 mmHg less due abnormal or non-oxygenated haemaglobin or methaemaglobin compounds and a further drop is due to the physiological shunt. This occurs at two sites – some blood from bronchial arteries branches of the descending thoracic aorta, returns via the pulmonary veins and left heart; and coronary artery blood flow may empty into the left atrial or ventricular chambers. The foetus, with high right-sided circulatory pressures, has the archetypal shunt via the ductus arteriosus.

7 **A** = False **B** = False **C** = True **D** = False **E** = True

Hypoxia, accumulation of CO_2 and a local rather than systemic fall in pH all cause pulmonary vasoconstriction. Pulmonary vessels receive sympathetic vasoconstrictor fibres unlike bronchioles, which although possessing an abundance of β_2-adrenoreceptors do not have an innervation to these and hence are predominantly sensitive to inhaled and circulating agonists only. Pulmonary veins are distensible and may contain a reservoir volume of up to 400 ml.

8 **A** = False **B** = False **C** = False **D** = True **E** = True

Average ventilation:perfusion ratio for the whole lung is 0.8 (4.2:5.5 litres/min) but at the apex there is relatively greater ventilation than perfusion. Physiological dead space is an excess of ventilation not equilibrating with blood. With exercise, cardiac output increases to 25–35 litres/min but maximal ventilatory volume may increase to 125–170 litres/min. Pulmonary embolism results in not only vascular obstruction but also alveolar collapse and haemorrhagic oedema to decrease ventilation. Haemorrhage causes veno- and vaso-constriction to reduce lung perfusion.

9 **A** = False **B** = True **C** = True **D** = False **E** = False

Increased H^+ concentration causing decreased oxygen affinity of haemoglobin is known as the Bohr effect and results from greater binding of deoxyhaemoglobin than oxyhaemoglobin to H^+ ion, hence causing a right shift of the curve to effect greater oxygen release. An increase in pCO_2 causes a concomitant decrease in pH. 2,3-dpg, a glycolysis metabolite, binds haemoglobin to release oxygen, thus causing a right shift, but note that acidosis reduces the amount of 2,3-dpg produced. Altitude and consequent increases in pH stimulate a substantial increase in 2,3-dpg. Abnormal haemoglobins have a lower P_{50} with greater affinity for oxygen and lesser release.

10 **A** = True **B** = False **C** = True **D** = False **E** = False

Each iron molecule binds one oxygen molecule as an oxygenation not oxidation reaction. It remains in the ferrous state. Binding and release of subsequent oxygen molecules changes the configuration of haemoglobin to give a sigmoid-shaped curve. Deoxyhaemoglobin has a greater affinity for H^+ and CO_2 than oxyhaemoglobin and results in facilitation for CO_2 transport and H^+ buffering in peripheral tissues. Carbonic anhydrase is contained in red cells and not in haemoglobin.

11 **A** = False **B** = False **C** = False **D** = True **E** = False

Haemoglobin affinity for carbon monoxide is extremely high and the latter is released very slowly. Carboxyhaemoglobin causes hypoxia as it causes a deficiency in the amount of haemoglobin able to carry oxygen, although total haemoglobin is unaltered. An additional problem is that the oxygen curve of the remaining oxyhaemoglobin is shifted to the left, further reducing oxygen release. Diagnosis is aided by measurement of carboxyhaemoglobin levels as pO_2 remains normal without concomitant respiratory stimulation. Treatment is the administration of 100% oxygen at increased barometric pressure to hasten carboxyhaemoglobin dissociation.

12 **A** = True **B** = False **C** = False **D** = False **E** = True

The main forms of CO_2 carriage involves its hydration and its formation of carbamino compounds with haemoglobin. CO_2 enters red cells and hydration occurs facilitated by carbonic anhydrase. The subsequent H^+ ions are buffered by haemoglobin and 70% of the HCO_3^- produced enters the plasma in exchange for Cl^- ions (the chloride shift). Little hydration occurs in plasma in the absence of carbonic anhydrase and HCO_3^- in plasma is predominantly originating from within the cell. Relatively little CO_2 is carried in the dissolved form and approximately 11% of total is carried as carbaminohaemoglobin.

13 **A** = False **B** = False **C** = True **D** = True **E** = True

CO_2 is 20 times more soluble than O_2 in plasma and more diffusible at the alveolo-capillary interface, the red cell membrane and the blood–brain barrier. Its hydration, although driven by carbonic anhydrase in red cells and in cerebrospinal fluid, is a reversible reaction. Unlike decreasing pO_2, increasing pCO_2 has a linear relationship with ventilatory stimulation. Changes in pO_2 will alter the slope of this relationship.

14 **A** = True **B** = True **C** = False **D** = True **E** = False

In polycythaemic states, the excess haemoglobin may not be fully saturated. In addition, polycythaemic red cells appear not to bind 2,3-dpg effectively thus impairing O_2 binding and release. It is primarily ventilation/perfusion inequality associated with pulmonary oedema which results in arterial hypoxaemia. Cyanosis may occur with a tension pneumothorax as it is a late sign and many other indicators of respiratory distress are emergent for this acute condition. Chronic airways disease as a result of hypoventilation and ventilation/perfusion inequalities, causes hypoxaemia and hypercapnia. It is not known why some patients respond with marked ventilatory drive with dyspnoea and others tolerate significant hypercapnia with impaired ventilatory control and little dyspnoea. Neonatal ventricular septal defects cause left to right shunting. Cyanosis will appear either late or if the shunt is severe, both leading to significant increases in right heart pressures and reversal of the shunt direction.

15 **A** = True **B** = True **C** = True **D** = False **E** = False

With severe head injury, damage to the respiratory centre may result either from direct axonal injury or with raised intracranial pressure and compression of medullary centres. Alkalosis associated with prolonged vomiting, attempts to inhibit ventilatory effort to effect CO_2 retention. This is not usually effective alone in returning the pH to normal without renal HCO_3^- loss and is compounded by hypoxia. In respiratory acidosis, the primary defect is CO_2 retention. pCO_2 is always elevated. Compensatory mechanisms include renal H^+ excretion and HCO_3^- retention which may or may not return pH to normal. Salicylates stimulate the respiratory centre. 'Restrictive' conditions cause hypoxia with relatively little change in pCO_2. Indeed pCO_2 may be low due to increased ventilation.

16 **A** = False **B** = False **C** = False **D** = True **E** = True

Increased pCO_2 beyond a normal range of 37–43 mmHg, increases HCO_3^- production in renal tubular and red blood cells. Chronic bronchitis elevates pCO_2 and lowers pO_2 but stimulation of ventilation may sufficiently return pCO_2 to normal. The remaining respiratory drive is thus hypoxic. Carotid and aortic afferent nerve section reduces response to CO_2 by only 30–40%, hence its effect on central chemoreceptors. CO_2 diffuses across the blood–brain barrier more readily than H^+, is hydrated and consequent rise in CSF $[H^+]$ stimulates medullary chemoreceptors. At altitude, reduced barometric pressure reduces O_2 tensions although air composition remains unchanged. Increased ventilation with a decline in pCO_2 occurs, causes respiratory alkalosis.

17 **A** = False **B** = True **C** = True **D** = True **E** = False

Lowered pCO_2 and $[H^+]$ ion moves the O_2 dissociation curve to the left. Respiratory alkalosis results in compensatory HCO_3^- excretion and also hypokalaemia as the relationship between $[H^+]$ and $[K^+]$ is reciprocal. In alkalosis, the hypokalaemia is due to a net increase in entry of K^+ into cells and increased urinary loss as H^+ is preferentially reabsorbed. A rise in pH causes reduction in liberation in free ionized Ca^{2+} from its salts in solution and also enhancement of calcium binding to plasma proteins; this tetany occurs due to a decreased ionized Ca^{2+} level with an unchanged total Ca^{2+} content.

18 **A** = True **B** = True **C** = False **D** = False **E** = False

Increased arterial pCO_2 stimulates the respiratory centre. It causes, via direct action, cutaneous and cerebral vasodilation, increasing cerebral blood flow to cause headache. Only insufficiently high arterial pCO_2 levels, e.g. with continuous inspired pCO_2 of 7%, have a narcotic effect with altered consciousness. The relationship between hypercapnia and cardiovascular regulation is complex. It has a direct stimulatory effect on the vasomotor area to cause tachycardia and peripheral vasoconstriction. However, stimulated peripheral chemoreceptor input on the vasomotor area causes bradycardia with peripheral vasoconstriction. Net effect is a tachycardia and vasoconstriction in most vascular beds except skin and brain.

19 **A** = True **B** = False **C** = True **D** = True **E** = True

Diffusibility of CO_2 exceeds that of O_2. Therefore with diffusion impairment which must be severe to be sufficient to cause hypoxaemia, ventilatory increases attempt to correct this, often with a slight lowering of pCO_2. It is a characteristic of a pure shunt, e.g. Fallot's tetralogy, that relatively little improvement is gained from increased inspired pO_2. Hyperventilation is a compensatory mechanism evoked by uraemia. A linear relationship exists between alveolar ventilation and arterial pCO_2.

20 **A** = True **B** = True **C** = False **D** = True **E** = True

With breech of lung or chest wall integrity, air moves down the passage of least resistance. Pulmonary deflation receptors are stimulated with increased vagal discharge. Tension pneumothorax is suspected by absent breath sounds/decreased air entry, hyperresonant percussion note, tracheal deviation away from the affected side and severe respiratory distress. It must be relieved immediately prior to X-rays. If severe, it is liable to cause kinking with impaired flow in intrathoracic great vessels and ultimately shock. With a closed pneumothorax, some thoracic cage volume is taken up by air in the pleural cavity.

21 **A** = False **B** = True **C** = True **D** = False **E** = False

Airway obstruction is most marked in expiration with decreased forced expiratory volume, and increased total lung capacity, functional residual capacity, and residual volume. Loss of lung elastic recoil allows unopposed chest wall elasticity to result in a barrel-shaped chest, contributing to increased lung volumes. Emphysematous destruction of lung tissue increases pulmonary vascular resistance with increase in right heart pressures and alveolar septal destruction causes confluence of alveoli. Hence larger but with fewer walls across which diffusion can occur.

22 **A** = True **B** = False **C** = True **D** = True **E** = True

Embolism ventilates underperfused lung segments (dead space). Also as a result of atelectasis, haemorrhage and oedema, perfusion of unventilated segments can occur (shunt). Oedema results in blood flow to unventilated units and contributes to ventilation/perfusion inequality as the main cause of hypoxaemia. Thickening of the alveolar–capillary blood–gas barrier and diffusion impairment appears to play a lesser role. Not only does alveolar oedema impair ventilation but airway resistance is typically increased also; this is due to oedema of the airways and reflex bronchoconstriction following stimulation of pulmonary irritant receptors.

23 **A** = False **B** = True **C** = True **D** = True **E** = False

Pain at the site of abdominal surgery is a particular problem in that it is exacerbated by movements such as respiration and coughing. Hence, in limiting the pain, the patient is at risk of underventilating with reduced tidal volume, atelectasis of such underventilated segments and lowering of arterial oxygen saturation. Abdominal surgery is less amenable to regional anaesthesia and although effective, opiates may compound the situation with respiratory depression.

24 **A** = True **B** = False **C** = False **D** = False **E** = False

Dissolved arterial oxygen alone is usually sufficient to meet the oxygen needs of the carotid and aortic bodies as their blood flow is so large. Haemorrhage of at least 15% total blood volume, that required to narrow the pulse pressure, reduces this flow to stimulate chemoreceptors and may also cause a stagnant hypoxia surrounding the chemosensitive cells. It decreases afferent discharge of carotid sinus baroceptors where as increased discharge occurs with greater perfusion pressures, to inhibit the respiratory centre. Stimulation of pulmonary inflation receptors serve to limit inspiration. 'Switching' between the phases of inspiration and expiration is automatic, controlled by ventral and dorsal medullary ganglia in the respiratory centre, and affected by pontine, cerebral cortex and vagal influences. Cerebral injury may disrupt any of these areas. Carbon monoxide poisoning does not significantly lower the arterial partial pressure of oxygen, as relatively little of the former is dissolved in plasma.

25 **A** = False **B** = True **C** = False **D** = False **E** = False

Anaemia reduces the total O_2 carriage in blood, but the remaining haemoglobin equilibrates normally with dissolved O_2. Hence there is no significant drop to 65 mmHg pO_2 sufficient to stimulate peripheral chemoreceptors. Hypoxaemia is a powerful stimulus but not until marked desaturation has occurred. Finer adjustments are effected by CO_2 and H^+ changes both on central and peripheral receptors and both will stimulate when elevated. High blood flow is sufficient for aortic and carotid body oxygenation from dissolved O_2 alone and increased flow will increase O_2 availability.

4 NUTRITION AND GASTROINTESTINAL FUNCTION

1 Regarding distribution of body protein, carbohydrate and electrolytes

A 40% of the total body Na^+ is in bone
B Body cell mass may be calculated from total body K^+
C Acidosis decreases free serum Ca^{2+}
D Total body glycogen weighs less than 1 kg
E Over 50% of body protein is intracellular

2 With reference to nutrition

A Total energy expenditure is similar pre- and post-operatively in elective minor surgery
B 5 grams of lean muscle tissue must be sacrificed to provide 1 gram of amino acid for gluconeogenesis
C 100 kcal per kg body weight are required daily by an average person
D 0.2–0.35 grams of nitrogen per gram body weight are required daily
E Urinary nitrogen approximates well to total nitrogen losses

3 Minerals and vitamins

A Niacin is a constituent of the flavoproteins
B Hypervitaminosis D causes weight loss and ectopic calcification
C Vitamin K is required to γ-carboxylate proline residues in clotting factors
D The daily requirement of folic acid is 10 mg
E Zinc is required to hydroxylate proline residues in collagen

4 Regarding the salivary gland

A pH of saliva is normally 7.4
B Normally 1500 ml of saliva is secreted daily
C The submandibular gland produces 70% of the daily output
D Potassium and bicarbonate occur in higher concentrations in the saliva than in plasma
E Amylase hydrolyses 1,6α linkages

5 Gastrointestinal fluid

A Gut secretions amount to approximately 7 litres per day
B Most fluid reabsorption occurs in the colon
C Gastrointestinal fluid loss is generally hypertonic
D Volume replacement with 5% dextrose is required for gut fluid loss or sequestration
E Ileostomy output is approximately 200 ml daily

6 Swallowing

A Is initiated by palatal elevation
B Is coordinated by the nucleus ambiguus and the tractus solitarius
C Is stimulated by glossopharyngeal nerve afferents
D The pharyngeal constrictors relax for food to pass through
E The oesophageal peristalsis wave takes around 6 s to reach the stomach

7 Stomach

A Mucus is secreted throughout the stomach
B Acid is secreted by glands in the fundus
C Volume is around 50 ml when fasted
D Has a positive respiratory quotient
E Has a pacemaker at the angula incisura

8 Stomach

A One to three litres of gastric juice are produced daily
B Secretion is stimulated by secretin
C Gastrin is produced by G-cells of the antrum at below pH 3
D Parietal cells have both gastrin and histamine receptors
E Mechanical stretching of the stomach wall will inhibit as the stomach becomes full

9 Regarding the duodenum

A Products of protein digestion and hydrogen ions cause gastric emptying when incident upon duodenal mucosa
B Is the major region of iron absorption
C Is largely responsible for folate absorption
D Has no significant net absorption of NaCl
E Does not secrete secretin

10 With regard to carbohydrate

A Glucose is absorbed primarily by active transport
B Insulin has almost no effect on gastrointestinal glucose absorption
C Maximum intestinal glucose absorption is approximately 120 g/h
D Maltose is absorbed after digestion to glucose and galactose
E Fructose is absorbed by facilitated diffusion

11 Regarding iron

A Twenty percent of ingested iron is normally absorbed
B Gastric pH reduces Fe^{2+} to Fe^{3+}
C Cereals may inhibit iron absorption
D It is stored in intestinal mucosal cells as mucosal transferrin
E Normal plasma iron is around 20 mmol/litre

12 The following are true with respect to protein digestion

A Pepsins hydrolyse bonds between aromatic and other amino acids
B Pepsins only function effectively in the stomach
C Trypsinogen is converted to trypsin by bicarbonate in pancreatic solution
D Only 80% of absorbed protein comes from ingested foods
E In infants moderate amounts of undigested protein are absorbed

13 In the colon

A Continence of the ileocaecal valve is necessary for efficient ileal absorption to occur
B Water is actively absorbed

C K^+ and HCO_3^- are secreted
D Clostridia are normal large bowel flora
E Large bowel flora produce useful vitamin K

14 In defaecation

A Mass movements require the integrity of the myenteric plexus
B The internal anal sphincter is inhibited by parasympathetic input
C The gastrocolic reflex is neurally mediated
D When rectal pressure reaches 55 mmHg the first urge to defaecate occurs
E Half the dry weight of faeces is bacteria

15 Regarding the gall bladder

A It has a ‘honeycomb’ mucosa
B It has Na^+ pumps in the mucosa
C It reduces the volume of bile stored
D It relies on the integrity of the sphincter of Oddi to allow it to fill
E When the bile duct is clamped gallbladder pressure takes several hours to reach 100 mm of bile

16 The following are cholagogues

A Fat
B Secretin
C $MgSO_4$
D CCK–PZ
E Gastrin

17 Gastrin

A Is secreted by cells of the APUD group
B May be found in foetal pancreatic islets
C Stimulates both insulin and glucagon secretion but only after a protein meal
D Atropine does not inhibit the gastrin response to a test meal
E Gastrin is inactivated in the kidney and the small intestine

18 Regarding bile

A Less than 1% of bile from the hepatic ducts is composed of bile salts and pigments
B Glucuronated (conjugated) bilirubin is water soluble and hence undergoes intestinal absorption
C Chenodeoxycholic acid is the most plentiful bile acid in humans
D Bile has a neutral pH before entry into the gall bladder
E Is essential for proper fat digestion

19 Bile

A Bile salts are sodium and potassium salts of bile acids
B Bile acids are tetrapyrrole derivatives
C The bile salt pool is around 300 mg
D Sodium glycocholate is absorbed in the ileum only by an active transport mechanism
E Enterohepatic circulation recycles the bile pigments six times daily

20 The following are true

A Unconjugated bilirubin is globulin bound in plasma
B Insulin increases bilirubin conjugation
C Conjugated bilirubin is renally excreted from plasma
D Icterus is clinically notable only at [bilirubin] > 35 mmol/litre
E Glucuronyl transferase in hepatocytes is bound to rough endoplasmic reticulum

21 Regarding fats

A Lipases are present in gastric chyme
B Lipids enter gastric cells by passive diffusion
C Free fatty acids travel to the liver in chylomicrons
D Cholesterol esters are found in all lipoproteins
E High density lipoprotein (HDL) is the main carrier of cholesterol to the tissues

22 The following stimuli increase gastrin secretion

A Non-cholinergic vagal discharge
B Blood calcium
C Adrenaline
D Blood secretin
E Blood insulin

23 Vasoactive intestinal peptide (VIP)

A Is found in nerves in the gastrointestinal tract
B Has a half-life in the blood of 10 s
C Stimulates intestinal secretion of electrolytes and water
D Inhibits gastric acid secretion
E Potentiates the action of acetylcholine on the salivary glands

24 Regarding gastric acid production

A Parietal cells produce approximately 150 mmol of HCl
B Parietal cells have an alkaline cytoplasm
C Carbonic anhydrase produces H^+ for excretion
d Parietal cells are stimulated by insulin
E Many mitochondria are present in parietal cells

25 Regarding the peritoneum

A It has a surface area of 10 m^2
B Peritoneal fluid has a protein concentration of <3 g/dl
C Normal peritoneal fluid white cell count is <3×10^9/litre
D Peritoneal fluid has no fibrinogen clotting system
E Peritoneal fluid circulates via the diaphragmatic lymphatics

1 **A** = True **B** = True **C** = False **D** = True **E** = True

Since the intracellular water concentration of K^+ (150 mmol/litre) is fairly constant, measurements of total body K^+ can be used to calculate body cell mass. Acidosis increases free ionized Ca^{2+} as plasma protein binds H^+ preferentially for conservation of pH.

2 **A** = True **B** = True **C** = False **D** = True **E** = True

Only in very major surgery is the energy expenditure raised, and then by up to 20–40%. This is similar to trauma and severe sepsis. Burns can double it. Tissue healing requires glucose for phagocytes and fibroblasts. In post-op starvation this requires gluconeogenesis, leading to significant lean tissue losses. In simple starvation, ketosis feeds the brain, and glucose requirements are much lower. One gram of adipose tissue provides nearly 1 gram of fat for metabolism. Only 50 kcal/kg/day are normally required; 150 kcal are required per gram of nitrogen. There are minimal nitrogen losses outside urinary urea (the end stage of the protein metabolism cycle) so E is true.

3 **A** = False **B** = True **C** = False **D** = False **E** = False

Riboflavin (B_2) is a constituent of the flavoproteins; deficiency causes glossitis and cheilitis. Niacin is a constituent of NAD and NADP deficiency causes pellagra (the four D's: dermatitis, diarrhoea, dementia and death). Vitamin D is converted in liver to 25-DHCC, then in kidney to 1,25-DHCC which enhances gastrointestinal Ca^{2+} and phosphate absorption. Clotting factors II, VII, IX and X require vitamin K for γ-carboxylation of glutamate residues. Recommended folate intake is 0.4 mg daily. Although zinc is required for tissue healing, it is ascorbic acid (vitamin C) which is needed to hydroxylate proline.

4 **A** = True **B** = True **C** = True **D** = True **E** = False

The pH of saliva is in the range 7.0–8.0, and is buffered by secreted HCO_3^-. The maximum salivary flow exceeds resting blood flow but on average 1500 ml of saliva is produced daily of which 25% is serous and produced by the parotid gland, 5% is mucinous and produced by the sublingual gland and 70% is of mixed type and produced by the submandibular gland. Na^+ and Cl^- are resorbed in the ducts whilst K^+ and HCO_3^- are secreted. Amylase hydrolyses at 1,4α linkages to oligosaccharides, limit dextrins, maltose and maltotriose. It works at optimum pH 6.7.

5 **A** = True **B** = False **C** = False **D** = False **E** = False

The small bowel is presented with approximately 9 litres of fluid daily, of which 2 litres are oral intake and 7 litres due to saliva, gastric and pancreatic juices, bile and duodenal secretions. About 5.5 litres are reabsorbed in the jejunum, the remainder passing on into the ileum, which removes a further 2 litres. The colon receives approximately 1.5 litres, most of which is reabsorbed leaving 200 ml in faeces. Gut losses are isotonic or slightly hypotonic and normal saline is required for replacement. Although adaption occurs, ileostomy output is usually that volume normally passing into the colon, 1–1.5 litres.

6 **A** = False **B** = True **C** = True **D** = False **E** = True

Swallowing is initiated by voluntary tongue movement pressing a bolus against the soft palate and the posterior pharynx. The afferents of cranial nerves V, IX and X stimulate the tractus solitarius and the nucleus ambiguus. Nerve IX may also stimulate the gag reflex. The reflex part of swallowing ensues, the soft palate closes and the larynx is then elevated. The bolus is squirted down the oesophagus by the powerful (100 mmHg) contraction of the constrictors. It is followed by a primary wave of peristalsis, travelling at 5 cm per second. If oesophageal stimulation continues a second peristaltic wave will ensue.

7 **A** = True **B** = False **C** = True **D** = False **E** = False

Mucus is secreted by neck cells of all glands of the stomach. HCl is secreted by parietal cells in the body of the stomach. A fasted stomach volume of 50 ml is normal. The respiratory quotient V_{CO}/V_{O} is negative in the stomach as acid production results in a net uptake of CO. A pacemaker for the slow gastric wave is triggered each 20 s by a pacemaker in the circular muscle of the fundus.

8 **A** = True **B** = False **C** = False **D** = True **E** = False

One to three litres is the volume of gastric juice secreted daily depending on source. Secretion is initiated by cranial nerve X (cephalic phase) directly and via gastrin cell stimulation. Food in the stomach stretches the walls and produces receptive relaxation and direct stimulation of the G-cells. Chemical content of food also stimulates G-cells. Gastrin and H receptors act synnergistically on parietal cells. Gastrin production is regulated via inhibition at pH < 3.0, in addition the small bowel hormones; secretin, VIP, GIP are released upon amino acids and low pH reaching small bowel mucosa, and inhibit gastric secretion.

9 **A** = False **B** = True **C** = False **D** = True **E** = False

Products of protein digestion and hydrogen ions stimulate duodenal production of CCK–PZ, secretin, VIP and GIP. They also stimulate the neurally mediated entrogastric reflex which slows gastric emptying. Iron, mainly as Fe^{2+} is absorbed largely in the duodenum. Folate absorption occurs mainly in the proximal jejunum. NaCl and H_2O are absorbed and secreted in similar quantities in the duodenum.

10 **A** = False **B** = True **C** = True **D** = False **E** = True

Glucose is absorbed by secondary active transport with a Na^{+}/glucose synport. Insulin has no significant effects on glucose uptake, which is maximal at around 120 g/h. Maltose is digested by brush border oligosaccharidases to glucose. Lactose is digested to glucose and galactose. Fructose is absorbed by facilitated diffusion and pentoses by simple diffusion.

11 **A** = False **B** = False **C** = True **D** = False **E** = True

Only 5% of ingested iron is absorbed. The majority is absorbed as Fe^{2+} but some is also absorbed as Fe^{3+} and haem itself. Gastric pH reduces Fe^{3+} to Fe^{2+}. Phytase in cereals reacts with iron to form insoluble compounds. Fe^{2+} is absorbed by binding to mucosal transferrin but is stored in mucosal cells bound to apoferritin as ferritin. The quantity of ferritin is inversely proportional to the amount of iron entering the blood stream. Concentrations are ~20 mmol/l in women with slightly higher levels in men.

12 **A** = True **B** = True **C** = False **D** = False **E** = True

Stomach pepsinogens are activated by low pH. Pepsins cleave at aromatic amino acids and have an optimal pH of 1.5–3.0, which means they are inactive outside the stomach. Pancreatic trypsinogen is activated by duodenal enterokinase and by activated trypsin. Active absorption occurs for L-isomer amino acids. Fifty percent of absorbed protein is derived from mucosal slough and digestive secretions. In infants IgA is absorbed in an undigested form.

13 **A** = False **B** = False **C** = True **D** = True **E** = False

In animals ileocaecal incontinence leads to rapid transit and malabsorption – this is not true for humans. Water is absorbed in the colon down an osmotic gradient produced by Na^+ absorption. Hypokalaemia and metabolic acidosis may be the sequelae of diarrhoea. Normal bowel flora are composed of amongst others; *E. coli*, *Strepfaecalis*, *Actinomycoses*, *Bacteroides* and *Clostridia*. Bacterial vitamin K is only usefully absorbed following coprophagia, normal behaviour for rats but not for humans.

14 **A** = True **B** = True **C** = False **D** = False **E** = False

In mass action large confluent areas of muscle contract with the effect of moving bowel contents and this requires an intact myenteric plexus. Sympathetic stimulation results in contraction of the anal sphincter and parasympathetic stimulation results in relaxation. The gastrocolic reflex is thought to be mediated by gastrointestinal hormones most probably gastrin. At 20 mmHg the first urge to defaecate is felt, at 55 mmHg defaecation is reflex. A quarter to a third of the dry weight of faeces is composed of bacteria.

15 **A** = True **B** = True **C** = True **D** = True **E** = True

The highly convoluted mucosa of the gall bladder gives it a classic macroscopic honeycomb appearance. The gall bladder reduces the volume of bile stored by absorbing Na^+ actively and with water following down the ensuing osmotic gradient. If the sphincter of Oddi is not closed then bile cannot 'back log' and fill the gall bladder. The clamped gall bladder will take several hours to reach this pressure of bile secondary to the on-going process of water resorption. With no gall bladder three times this pressure is reached within 30 min.

16 **A** = True **B** = False **C** = True **D** = True **E** = False

Cholagogues stimulate gall bladder contraction.

17 **A** = True **B** = True **C** = True **D** = True **E** = True

Gastrin is a polypeptide with at least three main forms: G_{14}, G_{17} and G_{34}. APUD (amine precursor uptake and decarboxylation) gastrin cells in the pyloric antrum secrete G_{17}. B helps to explain the origin of the pancreatic gastrinoma. The main effects of gastrin are stimulation of gastric acid and pepsin secretion. Lack of atropine inhibition of gastrin response to a test meal is due to the release of gastrin releasing peptide from the vagal nerves and not acetylcholine.

18 **A** = True **B** = False **C** = False **D** = False **E** = True

Duct bile is 97% water – this is concentrated five-fold in the gall bladder. Cholic acid composes the majority of human bile acid. Bile pH is 8.0 on entry into the gall bladder, which neutralizes and concentrates it. Around 25% of ingested fat will appear in the faeces in the absence of bile acids.

19 **A** = True **B** = False **C** = False **D** = True **E** = False

Bile salts are sodium and potassium salts of bile acids, which are cholesterol cyclopentanoperhydrophenanthrene derivatives. Three hundred mg are produced daily but the bile salt pool is 3–4 g. Bile acids are conjugated to glycine or taurine, hence glycocholic acid. The terminal ileum contains a specific transport mechanism for bile salts. This enterohepatic circulation recycles the bile salts six to eight times daily. Bilirubin is a tetrapyrrole porphyrin derivative.

20 **A** = False **B** = False **C** = True **D** = False **E** = False

Unconjugated bilirubin is bound to albumin in the plasma. Insulin has no effect on bile conjugation. Icterus is clinically notable at [bilirubin] > 35 mol/litre. Glucuronyl transferase is bound to the smooth endoplasmic reticulum of hepatocytes.

21 **A** = True **B** = True **C** = False **D** = True **E** = False

Lipases from the pharyngeal glands lead to digestion of 10–30% of fats in the stomach. Passive diffusion from micelles in the unstirred layer at the brush border allows lipids to enter stomach cells. Free fatty acids travel to the liver directly in the blood if short chain or are converted to triglyceride for transport in chylomicrons. Low density lipoprotein (LDL) is the main carrier of cholesterol to the tissues.

22 **A** = True **B** = True **C** = True **D** = False **E** = False

Vagal gastrin releasing peptide is the most likely transmitter and not acetylcholine.

23 **A** = True **B** = False **C** = True **D** = True **E** = True

VIP has a half-life in blood of 2 min.

24 **A** = True **B** = False **C** = False **D** = False **E** = True

An H^+/K^+ ATPase 'proton pump' transports H^+ out of the cell directly from ionized water produced within the cell. Carbonic anhydrase replaces the H^+ involved in this to 'regenerate' the water and simultaneously a HCO_3 leaves the cell from the interstitial side; creating balance. Insulin produces hypoglycaemia which causes central vagally mediated secretion of gastric acid. This is the basis of the test for the efficiency of vagotomy using insulin. Any cell which is secreting against a million-fold concentration gradient needs many mitochondria. The pH of parietal cell cytoplasm is 7.0–7.2.

25 **A** = False **B** = True **C** = True **D** = True **E** = True

The surface area of the peritoneum is approximately 1–2 m^2.

5 RENAL FUNCTION AND MICTURITION

1 Functional anatomy of the nephron

A There are approximately 3 million in each human kidney
B Nephrons are equally divided between cortex and medulla
C Intimal cells of the afferent arteriole form the macula densa
D A brush border is located upon epithelial cells of the distal convoluted tubule
E The efferent arteriole of each glomerulus provides a blood supply to neighbouring nephrons

2 Renal blood flow

A Is estimated using values of PAH clearance, extraction ratio and haematocrit
B Accounts for approximately 25% of cardiac output at rest
C Falls in hypovolaemic shock
D Is autoregulated largely through sympathetic innervation
E Arteriovenous O_2 difference is small compared to the heart

3 Regarding glomerular filtration

A GFR approximates 625 ml/min
B Inulin is freely filtered and actively secreted by tubules
C Anions are filtered slightly less readily than cations
D Plasma oncotic pressure rises as blood flows through the glomerulus
E Filtration fraction increases in shock

4 Transport mechanisms in the nephron

A Na^+ is actively transported out of the lumen of proximal tubules and into epithelial cells
B The tubular maximum for glucose reabsorption varies amongst nephrons of the same kidney
C Group specific symports exist for reabsorption of amino acids
D Clearance of PAH approaches that of inulin at high plasma concentration
E Tubuloglomerular feedback ensures a constant glomerular filtration rate

5 Water handling by the kidney

A Up to 70% of filtered water is reabsorbed in the proximal tubule
B The descending part of the loop of Henle is impermeable to water
C Fluid leaving the loop of Henle is hypotonic to plasma
D Cortical collecting ducts are impermeable to urea
E Some water is reabsorbed from the collecting ducts in the absence of ADH

6 Renal handling of an acid load

A H^+ is secreted in the proximal tubule in exchange for Na^+
B Aldosterone increases H^+ secretion in the distal tubule
C HCO_3^- is the main buffer throughout the nephron
D Maximal urine acidity is achieved at pH 3.5
E Low plasma HCO_3^- increases titratable acid in urine

7 Control of electrolyte reabsorption

A K^+ and H^+ compete for secretion in the distal tubule in exchange for Na^+
B A constant amount of filtered Na^+ is reabsorbed
C Cl^- and HCO_3^- are reabsorbed together in direct proportion
D Raised peritubular capillary hydrostatic pressure reduces Na^+ reabsorption
E Hyponatraemia develops with inappropriate ADH secretion

8 Diuretics and the kidney

A Hydrochlorothiazide reduces Na^+ reabsorption in the collecting ducts
B Acetozolamide increases excretion of Na^+, K^+ and HCO_3^-
C Amiloride antagonises vasopressin at the collecting ducts
D Spironolactone is of use in the treatment of hepatic oedema
E Frusemide reduces loop permeability to Na^+

9 Other drugs and the kidney

A Probenecid exacerbates gout
B An alkaline diuresis is a feature of salicylate overdosage
C Fludrocortisone exhibits minimal mineralocorticoid activity
D Methysergide typically causes renal parenchymal fibrosis
E Phenobarbitone clearance is impaired by renal disease

10 Acute renal failure

A Urinary Na^+ is raised in pre-renal oliguria
B Urinary casts accompany acute tubular necrosis
C Dangerous hypokalaemia may accompany uraemia and acidosis
D Urine:plasma urea ratio is low in intrinsic renal failure
E Progression to chronic failure is marked by a diuretic phase

11 Chronic renal failure

A Symptoms of uraemia develop at a blood urea level of 40 mmol/l or above
B A hypochromic microcytic anaemia is usually apparent
C Peritoneal dialysate containing 4 mmol/l K^+ accelerates K^+ clearance
D Inadequate elimination of water soluble drugs complicates haemodialysis
E Dialysis disequilibrium is related to plasma hypo-osmolality

12 The kidney as an endocrine organ

A Erythropoietin is produced by glomerular epithelial cells
B Erythropoietin stimulates aldosterone secretion
C Renin secretion is inhibited by angiotensin II
D 25-Hydroxylase deficiency leads to osteomalacia
E Local hormone secretion includes prostaglandin E2

13 Variations in renal physiology

A In a 70 kg male subject approximately 180 litres of fluid is filtered through the glomeruli each day
B Renal blood flow decreases with exercise
C Tubular maximum for glucose reabsorption increases with age
D Renin production is increased in shock
E High altitude stimulates erythropoietin production

14 Acidosis and hyperchloraemia occur in the following

- A Pyloric stenosis
- B Renal tubular acidosis
- C Use of carbonic anhydrase inhibitors
- D Ureteric implantation into the colon
- E Salicylate poisoning

15 Staghorn calculi

- A Are usually associated with infection
- B Are formed from calcium oxalate
- C Are more common in women
- D Predispose to intrarenal or perinephric abscess formation
- E Require treatment by surgical clearance and long-term antibiotics

16 The following are true of renal stones

- A They are commoner in Third World countries and probably related to dehydration
- B They are associated with anatomical abnormality such as medullary sponge kidney
- C Calcium oxalate stones account for the majority of renal calculi
- D Uric acid stones are usually radio-opaque
- E Most patients with renal stones have no detectable abnormality of urinary composition

17 Hypercalciuria is associated with the following

- A Secondary hyperparathyroidism
- B Prolonged immobilization
- C Myeloproliferative disorders
- D Sarcoidosis
- E Thalassaemia

18 The ureters and bladder filling

- A The ureters are innervated by both sympathetic and parasympathetic fibres forming an intramural plexus
- B A peristaltic wave passes along each ureter, the frequency of which is increased by sympathetic stimulation
- C Obstruction to ureteric drainage causes reflex renal arteriolar constriction
- D Intravesical pressure increases linearly with bladder filling
- E The first urge to void is felt at a bladder volume of about 500 ml

19 The micturition reflex

- A Is initiated by stretch receptors in the bladder wall via somatic afferents
- B Detrusor contraction is mediated by sympathetic efferents
- C Reflexes increase in rate and intensity with bladder filling
- D Stretch of the bladder neck causes reflex contraction of the sphincter urethrae muscle
- E Cervical spinal cord transection prevents the micturition reflex

20 Stored blood has the following characteristics

- A CPD-A preservative provides a shelf life of up to 35 days
- B One unit comprises of 500 ml of whole blood and a small amount of preservative
- C The O_2 affinity of Hb increases
- D Microaggregate formation is greater in red cell concentrates
- E Significant depletion of factors V, VIII and X

21 The following complications are associated with rapid transfusion of large volumes of blood

- A Disseminated intravascular coagulation
- B ARDS
- C A fall in ionized Ca^{2+}
- D Hypothermia
- E Jaundice

1 **A** = False **B** = False **C** = False **D** = False **E** = True

There are approximately 1.3 million nephrons in each kidney, of which 15% extend into the medulla. The macula densa is formed by the tubular cells at the junction between the loop of Henle and the distal convoluted tubule; juxtaglomerular cells are modified intimal cells of the afferent arteriole. A distinct brush border is found upon epithelial cells of the proximal convoluted tubule, reflecting its reabsorptive and secretory function.

2 **A** = True **B** = True **C** = True **D** = False **E** = True

Autoregulation is probably achieved by tonic contraction of afferent arteriolar vasculature in response to stretch. Acute hypotension below a systolic pressure of 80 mmHg results in renal vasoconstriction and reduced blood flow mediated by sympathetic nerves, circulating catecholamines and hypoxia. At low perfusion pressure glomerular filtration is maintained by efferent arteriolar vasoconstriction in response to angiotensin II. Because renal blood flow per gram of tissue is high, the O_2 arteriovenous difference is relatively low.

3 **A** = False **B** = False **C** = True **D** = True **E** = True

Inulin is used for measurement of GFR, which approximates to 125 ml/min, because it is freely filtered and neither secreted nor reabsorbed by the tubules. Whilst creatinine clearance is of value clinically, there is some active secretion and possibly reabsorption which affects results. Effective renal plasma flow as measured by PAH clearance is about 625 ml/min. Plasma oncotic pressure does indeed increase on passing through the glomerulus and anions are filtered marginally less readily than cations due to plasma proteins. Filtration fraction increases in shock because angiotensin II mediated efferent arteriolar vasoconstriction exceeds the reduction in glomerular blood flow.

4 **A** = False **B** = True **C** = True **D** = True **E** = True

Na^+ passively diffuses into proximal and distal tubule epithelial cells down a concentration gradient created by pumping it into lateral intercellular spaces. PAH is actively secreted in the proximal tubule. At plasma levels exceeding the tubular maximum for secretion, the amount filtered becomes more significant and clearance resembles that of inulin. Tubuloglomerular feedback maintains a constant rate of flow through the distal tubule. GFR is modified by constriction or dilatation of the afferent arteriole, probably mediated through renin-angiotensin, with cells of the macula densa acting as receptors for rate of Cl^- transport.

5 **A** = True **B** = False **C** = True **D** = True **E** = True

Cortical collecting ducts are impermeable to urea but permeable to water under the influence of ADH. As water is reabsorbed, urea concentration rises. In the medulla however, permeability increases and urea moves down a concentration gradient into the interstitium to support its osmolality. Reabsorption of urea by this means is flow dependent, most reabsorbed at low urine flow rate. Water is reabsorbed in the absence of ADH at the collecting ducts since a transtubular osmotic gradient always exists; a water diuresis can thus significantly dilute the interstitium.

6 **A** = True **B** = True **C** = False **D** = False **E** = True

Filtered HCO_3^- combines with secreted H^+ to form H_2CO_3. This is broken down by carbonic anhydrase on the brush border of proximal tubule epithelial cells to form H_2O and CO_2; the latter passes back into the cell and combines with H_2O to form H_2CO_3. This dissociates to form H^+, which is secreted, and HCO_3^- which is reabsorbed into the blood. NH_3 is produced in the distal tubule by the deamination of glutamine by glutaminase. It is an important buffer in the distal nephron along with HPO_4^{2-}, combining with H^+ to form NH_4^+. Maximal urine acidity is achieved at pH 4.5 which represents the limiting gradient against which transport mechanisms are able to secrete H^+. HCO_3^- is reabsorbed maximally at a plasma concentration of 28 mmol/l; above this level the urine is alkaline and below it H^+ appears as titratable acid and NH_4^+, even at alkaline pH ammonium ions will be detectable in the urine.

7 **A** = True **B** = False **C** = False **D** = True **E** = True

A constant proportion of filtered Na^+ is reabsorbed by a process of 'glomerulotubular balance', but control mechanisms exist in the distal segment of the nephron. Cl^- reabsorption is inversely proportional to HCO_3^- maintaining plasma anion concentration. Dilutional hyponatraemia occurs with inappropriate ADH secretion.

8 **A** = False **B** = True **C** = False **D** = True **E** = False

Thiazides reduce Na^+ and Cl^- co-transport at the distal tubule and increase flow-dependent secretion of K^+. Amiloride and triamterene interfere with Na^+ reabsorption in exchange for K^+ secretion at the distal portion of the distal tubule. Spironolactone inhibits the aldosterone-dependent Na^+ reabsorption at the collecting ducts. Acetazolamide inhibits carbonic anhydrase in the proximal tubule, decreasing H^+ secretion in exchange for Na^+ and reducing reabsorption of HCO_3^-. Because H^+ competes for secretion with K^+ by the Na^+ exchange pump, more K^+ is lost. Frusemide inhibits the $Na^+K^+2Cl^-$ co-transporter on the thick portion of the ascending loop.

9 **A** = False **B** = False **C** = False **D** = False **E** = False

Probenecid inhibits uric acid reabsorption at the proximal tubule and is used in the prophylaxis of gout. It also reduces tubular secretion of penicillins and certain cephalosporins. A forced diuresis with i.v. fluid and HCO_3^- facilitates renal H^+ excretion in the treatment of severe metabolic acidosis. Fludrocortisone has a high mineralocorticoid activity whilst that of betamethasone and dexamethasone is minimal. Methysergide may cause retroperitoneal fibrosis leading to obstructive renal failure. Water-soluble drugs such as gentamicin are cleared by glomerular filtration and tubular secretion; lipid soluble drugs generally depend upon hepatic elimination.

10 **A** = False **B** = True **C** = False **D** = True **E** = False

The urine profile of pre-renal oliguria resulting from impaired renal perfusion includes high osmolality, high urine:plasma urea ratio and low urinary Na^+ (less than 20 mmol/l). Inability to reabsorb Na^+ and secrete urea secondary to intrinsic renal disease results in low osmolality urine of high Na^+ content and low urea. Tubular blockage by 'Tamm–Horsefall' protein accompanies acute tubular necrosis and may appear as tubular casts in the urine. Recovery is generally heralded by onset of a diuretic phase.

11 **A** = False **B** = False **C** = False **D** = False **E** = True

Symptoms attributable to uraemia usually develop once the plasma level has exceeded 30 mmol/l. The anaemia of chronic renal failure is normochromic normocytic and results from lack of erythropoietin production. Acceleration of K^+ clearance by peritoneal dialysis is achieved by using a dialysate free of K^+. Haemodialysis removes water-soluble drugs (and vitamins) from the blood and may necessitate an added dose. It may also provoke 'dialysis disequilibrium' in which intracellular oedema develops owing to induced plasma hypo-osmolality.

12 **A** = True **B** = False **C** = False **D** = True **E** = False

Proximal tubular cells are the site for 1-hydroxylation, and hence activation, of 25-cholecalciferol, required for maintenance of Ca^{2+} homeostasis and the normal mineralization of bone.

13 **A** = True **B** = True **C** = True **D** = True **E** = True

14 **A** = False **B** = True **C** = True **D** = True **E** = False

$$[Na^+] + [K^+] = [HCO_3^-] + [Cl^-] + \underline{[X^-] + [A^-]}$$

The anion gap, single anion $[X^-]$ plus unmeasured anion $[A^-]$, is normally around 20 mmol/l. As $[Cl^-]$, $[X^-]$ and $[A^-]$ increase, $[HCO_3^-]$ falls. Acidosis develops as $[HCO_3^-]$:PCO_2 decreases. Metabolic acidosis therefore occurs with:

(1) Increased $[A^-]$. Secondary to either a reduction in GFR and hence HCO_3^- generation, or impaired acid secretion (carbonic anhydrase inhibitors and renal tubular defects) in which case K^+ is lost in exchange for Na^+ and Cl^- is reabsorbed instead of HCO_3^-.
(2) Increased $[Cl^-]$. Ureteric implantation into the colon facilitates Cl^- reabsorption in exchange for HCO_3^- by epithelial cells of the mucosa. RTA and CA inhibitors as above.
(3) Increased $[X^-]$. Salicylate stimulates an early respiratory alkalosis but also inhibits oxidative phosphorylation, promoting a lactic acidosis. Similarly ketones in keto acidosis. The vomiting of pyloric stenosis results in H^+ and Cl^- loss in excess of duodenal HCO_3^- culminating in a hypochloraemic alkalosis. H^+ secretion in the kidney is reduced and K^+ loss is accelerated.

15 **A** = True **B** = False **C** = True **D** = True **E** = True

Staghorn calculi are formed mainly from magnesium ammonium phoshate hexahydrate.

16 **A** = False **B** = True **C** = True **D** = False **E** = True

Most patients are idiopathic stone formers with 'high normal' levels of urinary calcium, urate, oxalate and glycosaminoglycans in combination with low urinary pH. Although dehydration is a risk factor, the highest incidence of renal calculi is in developed countries, probably related to overeating, with a male preponderance.

17 **A** = False **B** = True **C** = False **D** = True **E** = False

Primary and tertiary hyperparathyroidism are both causes of raised plasma calcium and hypercalciuria. Other common causes of hypercalciuria include carcinomatosis, myelomatosis, steroid treatment, hyperthyroidism, vitamin D overdosage and renal tubular acidosis. Myeloproliferative disorders predispose to uric acid stones.

18 **A** = True **B** = False **C** = True **D** = False **E** = False

Parasympathetic stimulation increases the frequency of peristaltic waves along the ureter.

The bladder wall exhibits 'plasticity' as it fills. There is a steady rise in intravesical pressure with up to 100 ml followed by a relative plateau phase punctuated by acute pressure increases or 'micturition waves'. At 400–500 ml pressure rises sharply; at this stage there is intense desire to micturate although the initial desire to do so is felt at about 150 ml.

19 **A** = True **B** = False **C** = True **D** = False **E** = False

Detrusor contraction is mediated via pelvic splanchnic (parasympathetic) nerves. Once bladder filling has occurred to an extent sufficient to stretch the bladder neck, there is a reflex relaxation of the sphincter urethrae to allow micturition.

Higher centres tonically inhibit the relex in addition to causing contraction of sphincter urethrae. Cord transection above the sacral segments removes cortical control but does not otherwise affect the micturition reflex, resulting in the 'automatic bladder'.

20 **A** = True **B** = False **C** = True **D** = False **E** = True

In each unit of whole blood there are approximately 430 ml of blood and 70 ml of anticoagulant/preservative (citrate phosphate dextrose and adenine). The addition of adenine helps to minimize the depletion of 2,3-dpg which increases the affinity of Hb for O_2 and hence reduces post-transfusional O_2 availability to the tissues.

Microaggregates are formed from platelets and granulocytes in whole blood preparations.

21 **A** = True **B** = True **C** = True **D** = True **E** = True

Massive transfusion (>8 units over 4 hours or >15 units over 24 hours) is complicated by DIC due to the reduction in coagulation factors and functioning platelets in stored blood and also to the activation of the coagulation cascade and complement by microaggregates. Microaggregates are also implicated in the pathogenesis of ARDS and a blood filter of 20–40 μm is recommended in addition to the transfusion of recently collected blood.

Citrate toxicity may produce a fall in ionized calcium and the infusion of calcium gluconate is occasionally required.

6 ENDOCRINOLOGY

1 The thyroid gland

A Descends from the pharyngeal roof along the thyroglossal duct
B Thyroid tissue is present in all vertebrates
C Has large follicles in the active state
D Usually maintains an iodine thyroid:plasma ratio around 10
E Synthesizes thyroglobulin within the colloid

2 T4

A Is proportionally more protein-bound than T3
B Has a long half-life compared to T3
C Is converted to tri-iodotryptophan in peripheral tissues
D Is more biologically active than T3
E Increases the number of LDL receptors

3 The adrenal medulla

A Is formed from ectoderm
B Is innervated by preganglionic cholinergic axons
C Secretes dopamine
D Is responsible for the majority of noradrenaline in serum
E Is responsible for the majority of adrenaline in serum

4 The following have high glucocorticoid:mineralocorticoid activity compared with cortisol

A Dexamethasone
B Prednisolone
C Corticosterone
D Deoxycortisone
E Aldosterone

5 Glucocorticoids increase numbers of circulating

A Eosinophils
B Neutrophils
C Basophils
D Platelets
E Erythrocytes

6 Cushing's syndrome

A May be iatrogenic
B May be of pituitary origin
C Is often of hypothalmic origin
D May be due to pituitary pathology
E May be nosocomial

7 Aldosterone

- A Causes natriuresis
- B Increases the sodium concentration of sweat
- C Secretion causes hypokalemia
- D Secretion increases response to circulating angiotensin II
- E Exposes mRNA by attaching to a nuclear receptor

8 Regarding the zona glomerulosa of adrenal cortex

- A Histologically comprises a network of cell columns
- B Produces cortisol
- C Produces mineralocorticoids
- D Is the major functioning zone in the fetus
- E Is attached to the renal capsule

9 PTH (parathyroid hormone)

- A Is derived from a precursor molecule in the anterior pituitary
- B Increases bone resorption of Ca^{2+}
- C Increases the formation of 1,25-DHCC
- D Directly increases uptake of Ca^{2+} from the gastrointestinal tract
- E Increases plasma phosphate levels

10 Calcitonin secretion

- A Is from the ultimobrachial bodies in the floor of the fourth ventricle
- B Directly inhibits osteoclasts
- C Increases phosphate resorption from bone
- D Increases calcium excretion in the proximal convoluted tubule
- E Is increased in Zollinger–Ellison syndrome

11 Regarding 1,25-DHCC

- A 1-Hydroxylation occurs in the liver
- B It mobilizes Ca^{2+} and phosphate from the bone
- C Directly increases Ca^{2+} and phosphate absorption from the gastrointestinal tract
- D Circulates mainly unbound in plasma
- E In excess may cause rickets in children

12 The following are secreted by eosinphilic cells of the anterior pituitary

- A GH
- B Prolactin
- C ACTH
- D Oxytocin
- E Lipotropin

13 The posterior pituitary

- A Shares an embryological origin with the intermediate lobe
- B Secretes neurophysin
- C Synthesizes ADH and oxytocin

D Consists of axons from the supraventricular and paraoptic nuclei of the hypothalmus
E Is a circumventricular organ

14 Growth hormone

A Resembles prolactin structurally
B Produces a positive nitrogen balance
C Raises blood sugar levels
D Levels are low in Ruwenzori pygmies
E In excess, causes gigantism in adults

15 ACTH (adrenocorticotrophic hormone)

A Is derived from pro-opiomelanocortin
B Has MSH activity
C Levels are low in congenital adrenal hyperplasia
D Is secreted in irregular bursts within an overall circadian rhythm
E Is always raised in Cushing's disease

16 Regarding prolactin secretion

A Physiological function in males is unknown
B It is lowered in Sheehan's syndrome
C Levels are increased by nipple stimulation
D It increases after parturition
E Hypersecretion is a cause of polymenorrhoea

17 Oxytocin

A Is increased by nipple stimulation
B Causes milk secretion in breast tissue
C Is increased by genital stimulation
D Causes sustained contraction of the gravid uterus
E In synthetic form is called ergometrine

18 The pineal gland

A Arises from the roof of the third ventricle
B Hassals corpuscles may calcify to give 'pineal sand' on X-ray
C Secretes MSH
D Involutes at puberty
E Hormone secretion is greatest in daylight hours

19 The hypothalmus

A Has vascular connections to the anterior pituitary via the hypothalamohypophyseal tract
B Has major neural connections with the limbic system
C Synthesizes ACTH
D Secretes glycoprotein releasing factor
E Contains osmoreceptors

20 ADH secretion increases in response to

A Increased plasma osmolality
B Increased ECF volume
C Angiotensin II
D Nicotine
E Standing

21 Juxtaglomerular cells (JG cells)

A Are in the walls of renal afferent arterioles
B Secretion is proportional to intra-arteriolar pressure
C Secrete angiotensin
D Their secretion is increased by sympathetic activity
E Their secretion is decreased in Goldblatt hypertension

22 ANP (atrial natriuretic peptide) secretion

A Increases with NaCl ingestion
B Increases as a direct result of atrial muscle stretching
C Is proportional to the CVP
D Raises blood pressure
E Increases ECF volume

23 Menopause

A Age of onset is steadily decreasing
B Results in anovulatory menstrual cycles
C Results in increased plasma LH levels
D Results in increased plasma oestrogen
E Results in increased plasma testosterone

24 FSH (follicle stimulating hormone)

A Is structurally related to LH, ACTH and TSH
B Stimulates testosterone production in Leydig cells
C Is responsible for early follicular growth
D The FSH surge is responsible for ovulation
E Is low in bald men

25 Testosterone

A Levels are similar in adult males and females
B Inhibits LH secretion
C Increases libido
D Is converted to the inactive dihydrotestosterone in target cells
E Decreases sebaceous secretions

26 Progesterone causes

A Myometrial relaxation
B Development of breast ducts
C Inhibition of LH secretion
D Raised alveolar pCO_2
E A fall in body temperature at ovulation

27 Oestrogen causes

A Increased amount of uterine muscle
B Increased secretion of TBG
C Spider naevi formation
D Cervical mucous arborization
E Spinnbarkeit less than 5 cm

28 On arriving in the High Andes

A Erythropoeitin secretion rises promptly
B Alveolar pO_2 falls
C Alveolar pCO_2 rises
D There is decreased 2,3-dpg in red blood cells
E Coffee is served colder

29 Whilst deep-sea diving in the Pacific

A Inhaled water causes a decreased plasma volume
B Ambient pressure increases by 760 mmHg every 32 m
C Pneumothorax may be caused by rapid descent
D Rapid ascent may cause nitrogen bubbles to form in tissues
E Treatment of decompression sickness is with diuretics

30 Islets of Langerhans

A Number about 1 million in the normal pancreas
B Venous blood drains into the hepatic portal vein
C Secretes the glycoprotein glucagon from A cells
D Synthesizes insulin within the ER of B cells
E Secretes somatomedin from D cells

31 Insulin

A Has an identical structure in all mammals
B Increases glucose uptake in muscle, adipose and kidney
C Glucose administered i.v. may cause hypokalaemia
D Increases fatty acid synthesis in adipose tissue
E secretion increases 100-fold following a meal

32 In insulin-dependent diabetes mellitus (IDDM)

A Intestinal absorption of glucose is increased
B Renal absorption capacity of glucose is decreased
C There is often osmotic diuresis
D Post-operative recovery is prolonged
E There is familial clustering

33 Glucagon

A Is degraded in the liver
B Secretion is stimulated by cortisol
C Is gluconeogenic
D Is glycogenolytic
E Is lipogenic

34 The following increase insulin secretion

A ACh
B Beta-blockers
C Gastrin
D Glucagon
E Thiazide diuretics

35 Features of Cushing's syndrome include

A Potassium depletion
B Keloid scarring
C Thick, dry skin
D Osteomalacia
E Diabetes mellitus

1 **A** = False **B** = True **C** = False **D** = False **E** = False

The embyrological thyroid originates from the pharyngeal floor. It has small follicles with columnar cells in the active state and large follicles with flat cells at rest. It maintains a iodine thyroid:plasma ratio in the order of 25 but this may be increased to 250 in iodine deficiency. Thyroglobulin is synthesized in thyroid cells and stored in colloid.

2 **A** = True **B** = True **C** = False **D** = False **E** = False

T4 is 99.98% protein bound compared to 99.5% (T3) and therefore has a longer half-life. Most of its action is via perpheral deiodination to tri-iodotyrosine (T3) which is 5–10 times as potent. It increases numbers of LDL receptors which lowers cholesterol levels.

3 **A** = True **B** = True **C** = True **D** = False **E** = True

The adrenal medulla arises from neural crest cells which are ectodermal in origin. Although it may be considered to be a sympathetic ganglion, the neurotransmitter involved is ACh. It synthesizes and secretes half the body's dopamine (the other half from sympathetic ganglia). After adrenalectomy, plasma Na^+ is largely unchanged (the majority originating from the sympathetic ganglia) but adrenaline levels fall essentially to zero.

4 **A** = True **B** = True **C** = False **D** = False **E** = False

The above are listed in order of glucocorticoid:mineralocorticoid activity. Dexamethasone has negligible mineralocorticoid activity, whereas aldosterone is the major mineralocorticoid

5 **A** = False **B** = True **C** = False **D** = True **E** = True

These are part of the anti-inflammatory and immunosuppressive roles of glucocorticoids. Other effects are decreased lymphocytes and regression of lymph node and thymus tissue. Inhibition of phospholipase A2 decreases formation of LT, TX, PG and PC and inhibits fibroblastic activity.

6 **A** = True **B** = True **C** = False **D** = True **E** = False

The majority of Cushingoid states are of iatrogenic origin. Other causes include hypersecretion of ACTH, often from pituitary adenomas but occasionally ectopically from bronchial carcinoma. Hypersecretion of CRH from the hypothalmus is rare. Nosocomial pertains to hospital acquired infections and is irrelevant in this case.

7 **A** = False **B** = False **C** = False **D** = True **E** = True

Aldosterone helps maintain ECF volume by absorption of Na^+ from urine, sweat, saliva and gastric juices. It also helps regulate K^+ levels by exchanging this for the Na^+ reabsorbed, so may cause hypokalaemia in excess. Polypeptide hormones (other than thyroid hormones) attach to receptors on cell membranes, as a steroid, aldosterone attaches to nuclear receptors to expose mRNA binding areas.

8 **A** = False **B** = True **C** = True **D** = False **E** = True

The glomerulosa cells are whorl like, giving way to columns in the zona fasciculata which branches into networks in the reticularis area. All three zonae produce cortisol, although only the zona glomerulosa has aldosterone receptors and can produce mineralocorticoids. Both zonae fasciculata and reticularis are ACTH dependent and secrete cortisol and sex hormones.

The foetal cortex consists mainly of a foetal zone which atrophies at birth. It forms part of the foetoplacental unit involved in secretion of oestrogens from the placenta.

9 **A** = False **B** = True **C** = True **D** = False **E** = False

PTH is synthesized from pre-pro PTH in the chief cells of the parathyroid glands. It increases serum calcium levels by resorption of Ca^{2+} from bone, from the distal convoluted tubule (DCT) in the kidney and by increasing formation of 1,25-DHCC. It is the latter which increases Ca^{2+} uptake from the gastrointestinal tract. PTH decreases phosphate levels by its phosphoric action in the DCT.

10 **A** = False **B** = True **C** = True **D** = False **E** = True

The ultimobrachial bodies are derived from the fifth branchial arches and in humans become distributed in the thyroid as parafollicular cells. Calcitonin lowers serum calcium and phospate levels by directly inhibiting osteoclastic activity and increasing calcium excretion in the DCT.

Gastrin, CCK and glucagon all stimulate calcitonin release (presumably in preparation for a calcium load), it is therefore elevated in the Z–E syndrome of hypergastrinaemia.

11 **A** = False **B** = True **C** = True **D** = False **E** = False

Vitamin D is 25-hydroxylated in the liver and 1-hydroxylated in the kidney to produce the active hormone. In conditions of reduced need, the kidney may switch to 24-hydroxylation to produce the less potent 1,24-DHCC. It increases serum [Ca^{2+}] and [phosphate] increasing resorption from bone and gut and increases calcium resorption from the kidney. It is a steroid hormone, so characteristically is highly protein bound in plasma. Deficiency may cause rickets in children or osteomalacia in adults.

12 **A** = True **B** = True **C** = False **D** = False **E** = False

Anterior pituitary secretions are:
(1) acidophils – GH, PRL (GP)
(2) basophils – FSH, LH, ACTH, TSH (FLAT)
(3) chromophobes – ?ACTH
Posterior pituitary: ADH, Oxytocin. Lipotropin is secreted by the same cells secreting ACTH, i.e. basophil corticotrophic cells.

13 **A** = False **B** = True **C** = False **D** = False **E** = True

The anterior and intermediate lobes of the pituitary arise from the embryological pharyngeal roof.

Neurophysins are part of the precursor molecules of ADH and oxytocin. These are synthesized in the PVN and SON and transported along the axons to be secreted from the posterior pituitary. The posterior pituitary, the area postrema, the subfornical organ and the organum vasculosum of the lamina terminalis are all circumventricular organs, being part of the CNS but outside the blood–brain barrier.

14 **A** = True **B** = True **C** = True **D** = False **E** = False

Both acidophil hormones GH and prolactin are structurally similar. Being primarily an anabolic hormome, GH produces a positive nitrogen balance and increases substrate accumulation in the blood (i.e. increases hepatic glucose output)

Pygmies from the Ruwenzori mountains in Zaire have normal GH levels, but decreased local IGF-1 around epiphysial plates, this being the usual transport vehicle for the local action of GH on bone. In excess it produces gigantism in children and acromegaly after epiphysial growth plate closure.

15 **A** = True **B** = True **C** = False **D** = True **E** = False

In the anterior pituitary, POMC is hydrolysed to ACTH, LPH and endorphin. In CAH, deficient enzymes for steroid synthesis in the adrenal cortex results in hyperstimulation by raised levels of ACTH. Increased levels with a normally functioning cortex will cause Cushing's disease; however, a primary increase in glucocorticoid levels (e.g. exogenous cortical adenoma) will depress levels of ACTH.

16 **A** = True **B** = False **C** = True **D** = False **E** = False

Pituitary secretion of prolactin is tonically inhibited by PIH (dopamine). Sheehan's syndrome is ischaemic necrosis of the pituitary stalk so reducing the tonic inhibition on prolactin. Prolactin and oestrogen are antagonistic; levels of both rise in pregnancy, but it is the rapid drop in oestrogen levels at parturition which allows the lactogenic effects of prolactin to occur. Similarly, hyperprolactinaemia inhibits the action of oestrogen on the ovary and is a cause of amenorrhoea.

17 **A** = True **B** = False **C** = True **D** = False **E** = False

Nipple stimulation in lactating women causes oxytocin and prolactin levels to rise. Prolactin causes milk secretion into ducts, whereas oxytocin causes the ejection reflex. In parturition, genital stimulation by the descending foetal head increases oxytocin output which facilitates labour by causing intermittent clonic uterine contractions. The synthetic form is syntocinon.

Ergometrine causes a single prolonged contraction and is routinely given immediately after delivery.

18 **A** = True **B** = False **C** = False **D** = True **E** = False

Hassals corpuscles are found in the thymus. The pineal gland is thought to function as a timing device to synchronize internal events with the light and dark cycle by secreting higher levels of melatonin in the hours of darkness.

19 **A** = False **B** = True **C** = False **D** = False **E** = True

The hypothalmus has neuroconnections to the posterior pituitary by the hypothalamohypophyseal tract and vascular connections to the anterior pituitary via the portal hypophyseal vessels. It is the site of synthesis of the anterior pituitary releasing factors (all oligopeptides) causing trophic hormone release, and of the posterior pituitary hormones oxytocin and ADH which are transported to that site for secretion. Osmoreceptors help control ADH secretions.

20 **A** = True **B** = False **C** = True **D** = True **E** = True

ADH promotes retention of water in excess of solute. It is therefore stimulated by hyperosmolality and decreased ECF volume.

21 **A** = True **B** = False **C** = False **D** = False **E** = False

The position of JG cells allows them to respond to decreased afferent arteriolar pressure by secreting renin. Renin converts angiotensinogen to angiotensin I which is then converted in the lungs to angiotensin II, a powerful pressor agent. Sympathetic activity increases blood pressure via this mechanism using receptors. Goldblatt demonstrated that clamping one renal artery activated the above systems, causing sustained hypertension.

22 **A** = True **B** = True **C** = True **D** = False **E** = False

Most of the actions of the ANP are antagonistic to ADH. It reduces Na^+ reabsorption at the kidney, so causing a salt diuresis which lowers blood pressure and ECF volume.

23 **A** = False **B** = False **C** = True **D** = False **E** = True

The age of menarche is decreasing and of menopause is increasing. Menopause can be thought of as end organ ovarian failure to produce sex hormones. As a result, feedback causes an increase in both trophic hormones causing raised testosterone levels.

24 **A** = False **B** = False **C** = True **D** = False **E** = False

FSH is a glycoprotein, sharing an identical subunit with LH and TSH. ACTH is a polypeptide derived from a much larger precursor. It acts on Sertoli cells in the testis to maintain spermatogenesis. LH stimulates the Leydig cells to secrete testosterone. Although there is an FSH surge at ovulation, the LH surge is thought to initiate ovulation. FSH has no effect on hair follicles.

25 **A** = False **B** = True **C** = True **D** = False **E** = False

Testosterone levels are in the order of 20 times higher in males. It has a negative feedback effect on LH (which stimulates its production). DHT is 100 times as potent as testosterone. It increases and thickens sebaceous secretions. This is the so-called 'nature's contraception' effect of pubertal acne in males!

26 **A** = True **B** = False **C** = True **D** = False **E** = False

Progesterone generally causes smooth muscle relaxation, including that of the uterus. Effects on breast tissue are development of alveoli and lobules. It stimulates respiration so lowering the pCO_2 and is responsible for the rise in body temperature at ovulation.

27 **A** = True **B** = True **C** = True **D** = True **E** = False

Oestrogen is responsible for uterine hypertophy in puberty and pregnancy, its lack in the menopause allows atrophy. Spider naevi, palmar erythema, gynaecomastia and testicular atrophy are well seen in advanced liver disease due to failure of oestrogen breakdown. When unopposed by progesterone it facilitates crystallization of NaCl in watery cervical mucous to form a fern-like pattern. Progesterone causes a thick cervical mucus to be secreted, a drop of which cannot be stretched more than 2–3 cm.

28 **A** = True **B** = True **C** = False **D** = False **E** = True

At altitude a decrease in partial pressure of oxygen results in reduced pO_2 in the alveoli, and compensatory hyperventilation reduces the alveolar pCO_2. Increased glycolytic energy production causes raised 2,3-dpg levels, shifting the O_2 dissociation curve to the right. This decreases O_2 making oxygen more available to tissues. Decreased atmospheric pressure allows water to boil at lower temperatures.

29 **A** = True **B** = False **C** = False **D** = True **E** = False

Inhaled hypertonic seawater draws water into the lungs from the vascular compartment. Hypotonic fresh water moves in the opposite direction. Pressure rises 1 atmosphere every 10 m (32 ft). Pneumothorax may be caused by the rapid expansion of air in the lungs due to decreasing ambient pressure during ascent. Treatment of decompression sickness is by recompression and then slow decompression.

30 **A** = True **B** = False **C** = False **D** = True **E** = False

There are 1–2 million islets in the pancreas and venous drainage takes the hormones directly to the liver, the major target organ, for maximal effect. A cells (20%) form the linear peptide glucagon; B cells (70%) form insulin; D cells form somatostatin; and F cells form pancreatic polypeptide.

31 **A** = False **B** = False **C** = True **D** = True **E** = False

Insulin has been described as 'the hormone of abundance' facilitating anabolism of complex molecules when substrates are plentiful. It does not cause glucose uptake by the kidney. It causes K^+ to enter cells, significantly so when given with glucose. Total insulin secreted is approximately 40 iu/day, basal levels are in the order of 1 iu/h with approximately 5 iu secreted in the hour after each meal.

32 **A** = False **B** = False **C** = True **D** = True **E** = True

Intestinal and renal absorption of glucose is unchanged. However, more glucose is presented to the renal tubules and may exceed its absorption capacity. This causes glucose to spill over into the urine, taking water with it. NIDDM is thought to be genetically determined, concordance rates in identical twins approaching 100%. Familial clustering is likely in IDDM, concordance rates being 50% in identical twins.

33 **A** = True **B** = True **C** = True **D** = True **E** = False

Glucagon is involved in breakdown of complex molecules to maintain the blood glucose level. It is therefore gluconeogenic, glycogenolytic, ketogenic and lipolytic.

34 **A** = True **B** = False **C** = True **D** = True **E** = False

Insulin secretion is increased by ACh (from the vagus nerve) and adrenergic agonists. Gastrin along with other intestinal hormones (GIP, secretin, CCK–PZ) also stimulate insulin secretion presumably in anticipation of glucose load. Thiazide diuretics decrease body K^+ which feeds back to lower insulin secretion so inhibiting K^+ transport into cells.

35 **A** = True **B** = False **C** = False **D** = False **E** = True

Cushing's syndrome is due to raised glucocorticoid levels, these have mineralocorticoid actions in excess. Amongst other glucocorticoid actions is immunosuppression, protein catabolism causing muscle atrophy, decreased wound healing and thin subcutaneous tissue. Osteoporosis may be a feature and the anti-insulin effect of glucocorticoids may precipitate diabetes.

7 PRINCIPLES OF FLUID DISTRIBUTION AND MOLECULAR BIOLOGY

1 Regarding fluid compartments within the body

- A The intracellular compartment is the largest
- B The intracellular fluid amounts to approximately 60% of body mass
- C The extracellular fluid comprises interstitial and plasma fluid
- D Plasma fluid amounts to approximately 8% of body mass
- E Interstitial fluid amounts to approximately 15% of body mass

2 Regarding fluid compartments within the body

- A Intra- and extracellular water totals 40 litres in a 70 kg male
- B Intracellular water amounts to 20 litres in the average person
- C Extracellular water amounts to 15 litres in the average person
- D Plasma amounts to 3.5 litres in the average person
- E The intracellular water volume is approximately double the extracellular water volume

3 Body water

- A Total body water accounts for 60% of total body weight
- B The water content of lean body mass decreases in catabolic states
- C Red cell water accounts for approximately 1.5 litres of ECF volume
- D ICF volume is maintained at approximately 28 litres despite variations in ECF volume.
- E Blood volume in children is approximately 7% of body weight

4 Measuring fluid spaces

- A Plasma volume may be measured by using any water-soluble substance
- B Plasma volume may be measured by radioactive mannitol dilution techniques
- C Plasma volume may be measured by radioactive albumin dilution techniques
- D Plasma volume may be measured by radioactive red cell dilution techniques
- E To measure plasma volume a substance must remain within the intravascular compartment without metabolism or excretion

5 Measuring fluid spaces

- A The extracellular fluid volume includes plasma, CSF, and interstitial fluids
- B Inulin is used to measure the extracellular fluid volume as it passes freely into the extracellular space but not into cells
- C Combining techniques of inulin dilution and albumin dilution may give a value for the interstitial fluid volume
- D Total body water volume is needed to calculate interstitial fluid volume
- E Total body water volume is needed to calculate intracellular fluid volume

6 Regarding the composition of intracellular fluid

A The total ionic concentrations are the same as interstitial fluid
B Bicarbonate ions are relatively fewer than in plasma
C The concentration of chloride ions is negligible
D The concentration of potassium ions is equivalent to that of sodium ions in the extracellular fluid
E The sodium ion concentration is 20% of that of plasma

7 Regarding the composition of interstitial fluid

A Chloride contributes over 80% of the negative ionic charges
B The concentration of sodium ions is equivalent to plasma
C The total ionic concentration is equivalent to plasma
D The protein content of interstitial fluid is much greater than that of plasma
E The protein content of interstitial fluid is much greater than that of intracellular fluid

8 Osmolality

A Refers to the number of moles of substance per unit volume of water
B When compared with plasma is referred to as tonicity
C Plasma osmolality may be calculated approximately by knowledge of the concentrations of the sodium, glucose and urea concentrations
D The plasma osmolality depends mostly upon the concentration of sodium ions
E High molecular weight substances exert a greater osmolality than low molecular weight substances

9 Regarding sodium and potassium ions

A The total body content of sodium ions is approximately 3000 mmols
B The total body content of potassium ions is approximately 3000 mmols
C Plasma potassium ions are influenced by plasma pH
D Increasing hydrogen ion concentrations within the plasma causes a reduction in plasma potassium ion concentration
E Over 95% of the total potassium ion content of the body is intracellular

10 Regarding the potassium ion

A Plasma content amounts to approximately 175 mmols
B Plasma potassium ion concentration mirrors plasma pH in its changes
C In alkalosis the potassium ion enters cells in exchange for hydrogen ions
D In acidosis the hydrogen ion enters cells in preference to potassium ions
E Plasma Potassium ion concentration is a poor marker for total body stores of the ion

11 DNA

A In the DNA chain the nucleotides adenine and guanine are bound together by hydrogen bonds
B In the DNA chain the nucleotides guanine and cytosine are bound together by hydrogen bonds
C Nucleotide bases are arranged in groups of three to code for each amino acid

D Genetic information is converted into amino acid chains by the process of transcription
E During the process of meiosis the genetic material is converted into a haploid state

12 The following terms are correct
A Aneuploidy: the amount of genetic information is other than a multiple of the normal haploid amount
B Transcription: the conversion of mRNA bases to the corresponding amino acid sequence
C Translation: the polypeptide sequencing resulting from ribosomal processing of mRNA
D Exon: the base sequences of DNA responsible for amino acid translation
E Intron: the base sequences of DNA responsible for regulating polypeptide translation

13 The synthesis of protein molecules
A Is brought about by three distinct processes: transcription, translation and post-translational modification
B The DNA sequence is read from the 5′ end
C tRNA is responsible for attaching the amino acids to the correct portion of mRNA on the ribosome
D Unlike mRNA, tRNA has four bases to code for each amino acid; with the addition of a guanosine nucleotide to each molecule
E The protein molecule is 'built' starting at its N-terminal end

14 Regarding the synthesis of proteins
A Multiple copies of amino acid sequences can be formed by polyribosomes
B mRNA must be remanufactured constantly as it is used up during translation
C tRNA uses complementary base sequences to mRNA to locate the correct position on the mRNA to locate its amino acid
D Unlike mRNA molecules the tRNA molecules are reused
E The first amino acid translated on the ribosome is isoleucine because the ribosome commences translation when it recognises the triplet coding for isoleucine

15 Regarding the synthesis of protein
A The amino acid sequence translated represents the ultimate protein molecule
B Amino acid sequences for more than one protein are produced from mRNA chains
C Most proteins produced by cells contain the original mRNA sequence of amino acids
D Hormones are unusual examples of proteins as they are cleaved from precursor peptides
E Hormone proteins may be reproduced up to six times in a precursor peptide

16 Regarding transport across cell membranes

A An ion's transport is enhanced by the relative negative charge within the cells
B Non-polar molecules, because of their lipid solubility, readily cross the cell membrane
C Dissolved gases pass easily across the cell membrane, despite being polar molecules
D Ease of diffusion in the case of non-polar molecules is proportional to their size
E Diffusion of water across the cell membrane is via special channels, as the lipid bilayer repels water

17 With reference to cell membrane permeability

A Permeability is greatest for small positively charged molecules
B Uncharged water soluble substances diffuse more easily across the cell membrane than charged molecules
C Cell membrane located protein molecules may act as channels and pumps for molecules to cross the cell membranes
D Channels may be opened by external ligands or voltage changes within the cell
E Facilitated diffusion is an energy-dependent process

18 The voltage-dependent channel

A Is typified by the sodium ion channel of excitable cells
B Is dependent upon the fact that the ion is positively charged
C Is responsible for the propagation of action potentials from nerve to muscle fibres
D May allow negatively charged ions to leave the cell
E Allows any positively charged molecule to enter the cell via the voltage-dependent channel

19 With reference to ion movements across cell membranes

A Sodium ion concentration within the cell is maintained by the balance between the forces of high extracellular concentration outside the cell and charge within the cell
B Chloride ion concentration within the cell is maintained by the balance between the extracellular high concentration, promoting influx and specific ion channels, expelling the ion
C Potassium ion concentration within the cell is maintained by the opposing forces of relative cellular high concentration, promoting efflux, and the electrical gradient preventing efflux
D The relative charge within the cell is dependent upon the relative concentrations of ions within and without the cell.
E The resting intracellular concentration of ions is altered by the presence of channels which allow preferential passage of ions

20 With regard to the equilibrium potentials for ions

A The equilibrium potential is the membrane potential in volts at which ionic efflux and influx is equal

B The equilibrium potential for sodium ions in the average mammalian cell is strongly negative
C The equilibrium potential for chloride can be calculated by using the Nernst equation
D The Nernst equation is only useful for calculating the equilibrium potential for chloride ions
E The Nernst equation does not apply to excitable cells which possess channels for influx or efflux of ions

21 With regard to ion channels

A The membrane Na–K–ATPase pump is responsible for the transport of three sodium ions into the cell, in return for two potassium ions out of the cell
B The Na–K–ATPase pump is responsible for maintaining the low intracellular sodium ion concentration
C The Na–K–ATPase pump is responsible for maintaining the relative negative charge of the inside of the cell with respect to its outside
D The Na–K–ATPase pump is responsible for the phase of repolarization in excitable cells
E The Na–K–ATPase pump is an example of a synport

22 The following second messengers and receptors are correct

A Acetylcholine causes a rise in intracellular calcium ion concentration
B 1-Receptor activation causes a rise in intracellular diacylglycerol
C 1-Receptor activation causes a fall in intracellular cAMP
D Nicotinic receptors have no second messenger associated with their activation.
E 2-Receptor activation causes a fall in intracellular cAMP

23 Of the second messenger systems within the cell

A Calcium ion concentration within the cell is increased purely by inward flux of the ion via voltage-dependent channels
B The calcium ions act by binding to cytoplasmic proteins thereby activating protein kinases
C Ligand-dependent calcium ion channels allow calcium ions to leave the cell during depolarization
D Inositol triphosphate increases intracellular calcium ion concentration
E cAMP produces its effects by phosphorylating protein kinase A, and activating this enzyme to phosphorylate other intracellular enzymes in a cascade

24 The second messenger system within cells

A Diacylglycerol activates protein kinase C which phosphorylates intracellular enzymes
B Inositol triphosphate is formed from phosphoinositol bisphosphate by the action of phospholipase C
C The endoplasmic reticulum is activated to release calcium ions by a rise in cAMP
D Is down-regulated by chronic stimulation of the receptor
E cAMP acts by phosphorylating intracellular protein kinase A and activating intracellular enzymes via a cascade

25 Regarding nerves in the peripheral nervous system

A Only myelinated nerves contain an insulating layer of Schwann cell membrane
B The axon conducts electrical charge easily without the aid of active processes
C The resting membrane potential is approximately –75 mV with respect to the exterior of the cell
D Action potentials may travel in a retrograde fashion along the axon from terminal to cell body
E Saltatory conduction refers to the propagation of an action potential along unmyelinated nerve fibres

26 Regarding the action potential in peripheral nerves

A The amplitude of the action potentials is proportional to the magnitude of the stimulus voltage
B The frequency of the action potentials is proportional to the frequency of the stimulus
C The action potential when measured by electrodes placed inside and outside the axon is biphasic in nature
D The measurement of the latent period between stimulation and the beginning of the action potential is proportional to the distance between the stimulating and recording electrodes
E The latent period of the nerve fibre is a constant value irrespective of the class of neurone

27 Regarding the phases of the action potential

A Initially the interior of the axon depolarizes slowly secondarily to passive external local changes in the membrane potential brought about by the stimulating electrode.
B 'Threshold potential' is a term referring to the individual voltage at which the axon will depolarize and is unique to each axon.
C Depolarization within the axon leads to the membrane potential of the axon becoming positive with regard to the exterior.
D The magnitude of the 'spike potential' is approximately +30 mV.
E The 'spike potential' reaches the membrane potential for sodium ions as the open ion channels allow free entry of the ion into the axon

28 Stimulation of an action potential

A Occurs only when a depolarizing voltage of sufficient magnitude is applied to the axon
B Requires higher voltages when the duration of the stimulus is very short
C Prolonged slowly rising voltages are ineffective as accommodation within the axon alters its ionic content
D Requires higher voltages if the neurone has very recently been stimulated to fire
E Is unaffected by external sodium ion concentrations.

1 **A** = True **B** = False **C** = True **D** = False **E** = True

Of the fluid compartments of the body the intracellular compartment is the largest. The intracellular fluid volume accounts for approximately 60% of the total body fluid but not of body mass. The total intravascular fluid amounts to approximately 80 ml per kilo mass but this also includes the cellular components.

2 **A** = True **B** = False **C** = True **D** = True **E** = True

While it is true that the intra- and extracellular fluid volume amounts to approximately 60% of body mass (40 litres), two-thirds of the fluid is intracellular. Therefore the volume of extracellular fluid is approximately 14 litres and the intracellular fluid 28 litres. Of the extracellular fluid three-quarters is interstitial and the remaining intravascular.

3 **A** = True **B** = False **C** = False **D** = False **E** = False

In catabolic states the proportion of water in lean body mass increases. Red cell water (1.5 litres) is intracellular; plasma volume accounts for approximately 3.5 litres of extracellular fluid. Blood volume in adults is roughly 7% of body weight; in children this increases to 8–9%. Loss of extracellular fluid leads to a redistribution of water throughout all compartments resulting in intracellular volume depletion.

4 **A** = False **B** = True **C** = True **D** = True **E** = True

To measure the plasma volume a substance which will remain within the vascular system without significant metabolism or excretion must be used. Substances which do not remain within the vascular compartment or are lost from the body give a falsely high plasma volume estimation.

5 **A** = True **B** = True **C** = True **D** = False **E** = True

The extracellular volume contains, by definition, all the fluid within the body but outside the cells. Extracellular fluid volume estimation requires the use of a substance which is freely permeable to the vascular endothelium but is not permeable to the cells themselves. Inulin is used for the measurement of extracellular fluid volume and albumin is used for the measurement of plasma volume.

6 **A** = False **B** = True **C** = True **D** = True **E** = False

The electrolyte composition of the two compartments are quite different with the intracellular compartment being hypertonic due to its high protein content. This accounts for the swelling which occurs in damaged cells. In the intracellular compartment the cation is predominantly potassium whereas in the interstitial fluid the predominant cation is sodium. The concentration of sodium ions in the intracellular fluid is less than 10% of the plasma.

7 **A** = False **B** = False **C** = False **D** = False **E** = False

The predominant cation in interstitial fluid is sodium, chloride ions are anions. The sodium ion concentration in the interstitial fluid is slightly less (~140 mmol/l) than that of plasma. In line with the relative reduction in the cation concentration of the interstitial fluid with respect to the plasma fluid so the total ionic concentration is also smaller. The interstitial fluid is virtually devoid of protein, whereas the plasma contains albumin and other proteins; the cells contain much protein in the form of enzymes and regulatory and structural protein molecules.

8 **A** = False **B** = True **C** = True **D** = True **E** = False

Osmolality is quite specific in its definition; it refers to the amount of substance per kilogram of solvent. Hypo- and hypertonicity refer to the comparison between the osmolality of a solution and that of plasma. A rough guide to the plasma osmolality may be calculated by adding twice the sodium ion concentration to twice the potassium ion concentration and adding the urea concentration and glucose concentration. The osmolality depends upon the total number of particles in a solution and not on the molecular weight of the substance.

9 **A** = True **B** = True **C** = True **D** = False **E** = False

The total body stores of each cation is approximately 3000 mmol, the difference being that sodium ions are mainly extracellular in location, and the potassium ions are mainly intracellular. Potassium ion and hydrogen ion concentrations are interrelated in the fact that they rise and fall together. This is due to the influx of hydrogen ions into the cell in exchange for efflux of potassium ions out of the cell in periods of high hydrogen ion concentration, and vice versa. Of the 3000 mmol of potassium ions in the body about 2700 mmol is intracellular.

10 **A** = False **B** = False **C** = True **D** = True **E** = True

Taking the plasma concentration of plasma as 4.5 mmol/l and the plasma volume as 3.5 litres then the total plasma potassium ion content is approximately 16 mmol. Hydrogen ion and potassium ion concentrations tend to rise and fall together, therefore a rise in pH will tend to lead to a fall in potassium ions as the pH represents $-\log_{10}[H^+]$. The fact that approximately 0.5% of the total body potassium ion content is contained in the plasma means that the plasma potassium ion concentration is a poor marker of total body stores.

11 **A** = False **B** = False **C** = True **D** = False **E** = True

The pairing nucleotides in the DNA molecule are adenine with thymine and cytosine with guanine. The amino acid are coded for by three base-pair nucleotides called a 'triplet'. Amino acid sequences are formed from the DNA code by the processes of translation and transcription. 'Haploid' refers to a genetic content exactly half of the normal 26 pairs of chromosomes.

12 **A** = True **B** = False **C** = True **D** = True **E** = False

Aneuploidy refers to a chromosome content within a cell which is other than an exact multiple of the haploid number (26) , this is a characteristic of malignant cells. Translation is the process in which mRNA bases are converted into amino acids; transcription, on the other hand, is the conversion of DNA bases to the corresponding mRNA bases. Exons are the portions of DNA which code for protein synthesis, whereas introns are regions of the DNA which are not translated.

13 **A** = False **B** = True **C** = True **D** = False **E** = True

There are four processes involved in protein synthesis; transcription, post-transcriptional modification, translation and post-translational modification. mRNA has the same 'triplet' system of coding for amino acids as does DNA, however using complimentary bases and the substitution of the base uridine in place of the thymine.

14 **A** = True **B** = False **C** = True **D** = False **E** = False

More than ribosome may, and often does, attach itself to the mRNA chain to produce multiple copies of the amino acid sequence. Transcription has no effect on the integrity of the mRNA and several copies of the mRNA's amino acid sequence are manufactured. tRNA is recycled similar to mRNA is protein synthesis. The first amino acid to be translated from the mRNA chain is methionine, as its coding 'triplet' is recognized by the ribosome.

15 **A** = False **B** = True **C** = False **D** = False **E** = True

The amino acid sequence formed by translation of the mRNA chain is subjected to post-translational modification which may yield several polypeptides, as in the case of hormones. Modification of the mRNA leads to the removal of bases from the original mRNA sequence. All amino acid sequences derived from mRNA translation are modified before the ultimate protein is produced. Multiple copies of the same protein molecule may be represented in the mRNA sequence produced, allowing, along with polyribosomal translation, multiple copies of the protein molecule to be produced from a single mRNA strand.

16 **A** = False **B** = True **C** = False **D** = True **E** = False

Anions are negatively charged ions, being drawn to the anode, and hence would be repelled from the relatively negatively charged cell interior. Small non-polar molecules cross the cell membrane with greater ease than similarly charged molecules of larger size and polar molecules cross the cell membrane with greatest difficulty. Dissolved gas molecules do not ionize and hence are not polar molecules. The lipid bilayer is highly permeable to water, obviating the need for channels.

17 **A** = False **B** = True **C** = True **D** = True **E** = False

Cell membrane permeability is greatest for small non-polar molecules, it is relatively impermeable to polar molecules. The nature of the charge is a relatively minor consideration. The cell membrane channels and pumps are protein molecules which pass several times across the cell membrane. The channels may be opened by voltage changes in the cell or by the actions of ligands. These ligands may extracellular as in the case of neurotransmitters, or internal in the cases of second messengers such as cAMP. Facilitated diffusion moves molecules across cell membranes in the direction of decreasing chemical gradients, and as such do not require the expenditure of energy.

18 **A** = True **B** = True **C** = False **D** = True **E** = True

The wave of depolarization which passes through excitable tissues, such as muscle fibres and neurones is brought about by the influx of sodium ions through voltage-dependent channels. The relatively negative voltage of the cell interior and lower internal concentration of the ion enhance this influx. The propagation of impulses from neurone to muscle fibre is produced by the release of acetylcholine from the nerve ending. The channel is capable of allowing negatively charged ions to leave the cell if their intracellular concentration is higher than the interstitium. It is obvious that if the molecule is larger than the channel then it cannot enter the cell by this route, irrespective of its charge.

19 **A** = False **B** = False **C** = True **D** = True **E** = False

The relative negative charge within the cell tends to draw the positively charged sodium ion into the cell. The intracellular concentration of chloride ions is negligible, this is due to the relative negative charge within the cell repelling the ion. The balance of ionic concentrations inside and outside the cell is the basis of the Nernst equation. In the resting state, as its name suggests, the cell is in a stable state with the rate of ionic movement into and out of the cell being matched. In a stimulated state the cell behaves differently and this is the basis of the cell's response to the stimulation.

20 **A** = True **B** = False **C** = True **D** = False **E** = False

The Nernst equation calculates, using the relative intra- and extracellular concentrations of the ion in question, the cell membrane potential at which the efflux and influx of the ion is matched. Sodium ions tend to be drawn into a cell by the combined high extracellular concentration and relatively negative intracellular charge, this means that a positive intracellular charge is needed to repel the in out of the cell against its concentration gradient. The Nernst equation stands true for all ions found inside the cell, as it does for all cells in an unexcited state. Stimulation of the cell to open ion channels allows ions to pass down their concentration gradients, altering the ionic composition of the cell.

21 **A** = False **B** = True **C** = True **D** = True **E** = False

The cell membrane Na–K–ATPase pump is a long chain protein molecule which passes back and forth through the cell membrane. By changing its configuration it can transport three sodium ions out of the cell in return for transporting two potassium ions into the cell, thereby maintaining the electrochemical balance of the cell. During depolarization, sodium ions flow into the cell causing a net positive charge to enter the cell, this is reversed by the pump which transports a greater number of sodium ions out than potassium ions into the cell, so restoring the resting membrane potential. A synport transports two different ions, or molecules, in the same direction across the cell membrane; this pump is an example of an antiport.

22 **A** = False **B** = True **C** = False **D** = True **E** = True

Acetylcholine receptors respond to ligand stimulation by the opening or closing of ion channels. 1-Receptors respond by activation of the diacylglycerol and inositol triphosphate pathways. cAMP concentrations are increased by the stimulation of 1-Receptors and decreased by the stimulation of 2-Receptors. Nicotinic receptors are stimulated by acetylcholine, and hence, respond by the opening or closing of membrane-bound ion channels.

23 **A** = False **B** = True **C** = False **D** = False **E** = True

Although it is true that stimulation of an excitable cell leads to the influx of calcium ions from the interstitial fluid, this is a minor consideration compared to the release of calcium ions from within intracellular stores. One of the actions of the released calcium ions is to bind to calmodulin and, thereby, activate myosin light chain kinase; in actively stimulated muscle fibres. During depolarization calcium ions enter the cell via channels, they are removed via an antiport in exchange for sodium ions or stored in the endoplasmic reticulum. Inositol triphosphate increases calcium ion concentrations by stimulating the release of calcium ions from their intracellular stores.

24 **A** = True **B** = True **C** = False **D** = False **E** = True

Activation of the diacylglycerol and inositol triphosphate second messenger system, by, for example, the 1-Receptor, leads to calcium ion release via the inositol triphosphate and the diacylglycerol activates protein kinase C to activate cellular enzymes. cAMP, on the other hand, leads to the activation of protein kinase A. Chronic stimulation of a receptor leads to down regulation of the receptor itself, which is the phenomenon of tachyphylaxis.

25 **A** = False **B** = False **C** = True **D** = True **E** = False

Both myelinated and unmyelinated nerves possess a outer covering of myelin produced by Schwann cells; the difference being in that in the myelinated nerve cell the myelin in wrapped around the fibre rather than just providing a covering. The nerve fibre, itself, is a very poor conductor of electrical charge passively. An electrical stimulation applied to the axon of a nerve fibre will lead to a wave of depolarization passing in both directions from the site of stimulation. The retrograde depolarization will not, however, pass across synapses as the neurotransmitters are located on the 'wrong' side of the synapse for the action potential to make use of.

26 **A** = False **B** = False **C** = True **D** = True **E** = False

Above the threshold voltage required to produce an action potential, no further increase in the voltage will increase the size of the action potential; this is the basis of the 'all or none' phenomenon of the neurone. While it is true that the rate of action potential production matches that of stimulation, when the voltage applied is greater than that of the threshold, this only applies to frequencies which do not exceed the neurone's ability to repolarize itself. Although conventionally regarded as monophasic in nature there is hyperpolarization of the neurone during the phase of repolarization. Latent times are proportional to both the distance between the stimulating and receiving electrodes and to the velocity of conduction. This velocity depends upon the size of the neuronal diameter and its myelination, or not.

27 **A** = True **B** = False **C** = True **D** = True **E** = False

Stimulation of the axon with positive charge causes local depolarization of the cell membrane. If this local depolarization is sufficiently large it leads to depolarization of the cell locally and production of an action potential. The 'threshold potential' is not unique to different classes of neurone and amounts to approximately 15 mV of depolarization. Activation of the neurone to produce an action potential leads to the opening of sodium ion channels in the cell membrane. This leads to the accumulation of positive charges within the cell and the membrane potential becoming positive. The influx of sodium ions causes the membrane potential of the neurone to approach that of the sodium ion; repolarization processes come into action at this point and the membrane potential never actually reaches that of sodium ions, peaking at +30 mV.

28 **A** = False **B** = False **C** = True **D** = False **E** = True

Stimulation of an action potential in an axon will not occur if a voltage of suitable proportion is applied to the neurone but the rate of change of the voltage is too slow. Slow depolarizing voltages result in homeostatic changes in the ionic composition of the cell to minimize the change in the cell membrane potential. This is called 'accommodation'. Very short duration voltages do not cause depolarization in the neurone because they are of insufficient duration to allow significant ion influx. After depolarization there is a period of absolute refractoriness in which the neurone cannot be made to produce an action potential irrespective of the voltage applied. The relatively small number of sodium ions needed to move into the neurone to produce depolarization means that the extracellular sodium ion concentration is relatively unimportant.